SpringerBriefs in Optimization

Series Editors

Panos M. Pardalos
János D. Pintér
Stephen M. Robinson
Tamás Terlaky
My T. Thai

SpringerBriefs in Optimization showcases algorithmic and theoretical techniques, case studies, and applications within the broad-based field of optimization. Manuscripts related to the ever-growing applications of optimization in applied mathematics, engineering, medicine, economics, and other applied sciences are encouraged.

For further volumes:
http://www.springer.com/series/8918

Anna Nagurney • Min Yu • Amir H. Masoumi
Ladimer S. Nagurney

Networks Against Time

Supply Chain Analytics for Perishable Products

Anna Nagurney
Isenberg School of Management
University of Massachusetts
Amherst, MA, USA

Amir H. Masoumi
Isenberg School of Management
University of Massachusetts
Amherst, MA, USA

Min Yu
Pamplin School of Business
University of Portland
Portland, OR, USA

Ladimer S. Nagurney
College of Engineering, Technology
 and Architecture
University of Hartford
West Hartford, CT, USA

ISSN 2190-8354 ISSN 2191-575X (electronic)
ISBN 978-1-4614-6276-7 e-ISBN 978-1-4614-6277-4 (eBook)
DOI 10.1007/978-1-4614-6277-4
Springer New York Heidelberg Dordrecht London

Library of Congress Control Number: 2012956441

Mathematics Subject Classification (2010): 90Bxx, 90B06, 90B10, 90B15, 90C30, 90C35, 91A10, 91A80, 91B55, 91B72, 58E35, 65K15, 90C33

© Anna Nagurney, Min Yu, Amir H. Masoumi, Ladimer S. Nagurney 2013

This work is subject to copyright. All rights are reserved by the Publisher, whether the whole or part of the material is concerned, specifically the rights of translation, reprinting, reuse of illustrations, recitation, broadcasting, reproduction on microfilms or in any other physical way, and transmission or information storage and retrieval, electronic adaptation, computer software, or by similar or dissimilar methodology now known or hereafter developed. Exempted from this legal reservation are brief excerpts in connection with reviews or scholarly analysis or material supplied specifically for the purpose of being entered and executed on a computer system, for exclusive use by the purchaser of the work. Duplication of this publication or parts thereof is permitted only under the provisions of the Copyright Law of the Publisher's location, in its current version, and permission for use must always be obtained from Springer. Permissions for use may be obtained through RightsLink at the Copyright Clearance Center. Violations are liable to prosecution under the respective Copyright Law.

The use of general descriptive names, registered names, trademarks, service marks, etc. in this publication does not imply, even in the absence of a specific statement, that such names are exempt from the relevant protective laws and regulations and therefore free for general use.

While the advice and information in this book are believed to be true and accurate at the date of publication, neither the authors nor the editors nor the publisher can accept any legal responsibility for any errors or omissions that may be made. The publisher makes no warranty, express or implied, with respect to the material contained herein.

Printed on acid-free paper

Springer is part of Springer Science+Business Media (www.springer.com)

Preface

Today, the nature of production, transportation, and consumption of numerous products, as well as services, is global. Products are routinely shipped across countries, continents, and hemispheres. This is rather paradoxical, since a wide range of products, from fresh produce to pharmaceuticals, radioisotopes, fast fashion, and even computers and consumer electronics, are time-sensitive and perishable. Hence, their supply chains, consisting of networks of economic activities, must be optimized accordingly to ensure that the products that consumers demand are delivered where and when they are needed, in such a way that losses and waste are minimized.

Time manifests itself in various forms in supply chain networks. Such activities as manufacturing, transportation, storage, and distribution to retailers and consumers clearly have time associated with them. It takes time to produce and ship a high technology product from the Far East to North America. It takes time to manufacture and package a drug. It takes time to transport fruits and vegetables across the equator so that customers have produce year-round. A day's production of a radioisotope must be used within a week in a hospital or medical facility, due to radioactive decay. Medicines and blood for transfusions that are not available in a timely manner may adversely affect the very survival of patients. Delays in executing supply chain network activities, especially in the case of time-sensitive and perishable products, may impact both the quantity and the quality of products available at demand markets, affecting not only the consumers but also the very profitability of a firm and/or the reputation of the organization.

Time is also associated with the dynamics of adjustment processes. Firms must learn about their competitors' behavior as well as consumers' preferences and then adjust their types and quantities of products. Clearly, consumers are also time-sensitive when it comes to the purchase of specific products. Hence, the inclusion of time elements into supply chain analytics is critical, especially in today's global scenario in which not only are decision-makers personally pressed for time but the same also holds for their organizational and business processes and products. Analytics, the scientific process of transforming data into insight for making better decisions, provides the platform for such prescriptive decision-making.

Indeed, decision-makers, whether individuals, firms, or other organizations, who optimize and compete with and against time, will not only have the full advantages of the best allocation of their constrained resources but will also achieve their desired objectives.

The goal of this book is to provide a unified framework for supply chain network analytics for perishable products. It emphasizes the underlying application-specific supply chain network structure, the behavior of the decision-makers, and the crucial time element. This book synthesizes a collection of research on this topic, focusing on generalized nonlinear networks, in the context of both optimization and game theoretic mathematical models. Through the use of case studies, this book explores such timely topics as blood supply chains, medical nuclear supply chains, fresh produce supply chains, pharmaceutical supply chains, and fast fashion apparel chains. It vividly depicts the various processes and provides an integrated variational inequality theory approach, which combines network theory, optimization, and game theory.

Following the Introduction and Scope in Part I, this book, in Part II, describes two distinct supply chain network optimization models and applications. In Part III, it turns to supply chain network competition. Models are constructed with different demand structures, uncertain, fixed, or elastic (price-sensitive). On the supply side, decision-makers' multiple criteria are identified to bring a richness to the decision-making possibilities. The theoretical foundations are laid out, and algorithms are constructed, which are then applied to distinct scenarios in a plethora of application-driven supply chain network case studies. The book reports the input data, how it was obtained, along with the computationally obtained output data and information, for transparency and clarity purposes.

This book is an outgrowth of our collaborations through the Virtual Center for Supernetworks at the Isenberg School of Management at the University of Massachusetts Amherst on a series of projects and papers. The first author gratefully acknowledges support from the John F. Smith Memorial Fund at the Isenberg School of Management and the support provided by the School of Business, Economics and Law at the University of Gothenburg in Sweden, which appointed her a 2012–2013 Visiting Professor. The research of the first two authors was also supported, in part, by the National Science Foundation (NSF) grant CISE #1111276, for the NeTS: Large: Collaborative Research: Network Innovation Through Choice project awarded to the University of Massachusetts Amherst with the first author as a Co-PI.

Amherst, MA	Anna Nagurney
Portland, OR	Min Yu
Amherst, MA	Amir H. Masoumi
West Hartford, CT	Ladimer S. Nagurney

Contents

Part I Introduction and Scope

1   Introduction and Scope 3
   1.1   Background 3
   1.2   The Methodological Framework 5
   1.3   The Organization of the Book 7
   References 8

Part II Supply Chain Network Optimization Models and Applications

2   Blood Supply Chains 11
   2.1   Motivation and Overview 11
   2.2   The Blood Supply Chain Network Model 13
       2.2.1   The Components of a Blood Banking System 13
       2.2.2   The Formulation 15
       2.2.3   Illustrative Blood Supply Chain Network Examples .. 25
   2.3   The Algorithm 29
       2.3.1   Explicit Formulae for the Euler Method Applied
               to the Blood Supply Chain Network Variational
               Inequality (2.38) 30
   2.4   A Case Study 31
   2.5   Summary and Conclusions 34
   2.6   Sources and Notes 34
   References 35

3   Medical Nuclear Supply Chains 37
   3.1   Motivation and Overview 37
   3.2   The Medical Nuclear Supply Chain Network Model 39
       3.2.1   The Underlying Physics 43
       3.2.2   The Multicriteria Decision-Making Problem in Link
               Flows and in Path Flows 46

	3.3	The Computational Approach 48
		3.3.1 Explicit Formulae for the Lagrange Multipliers at Step 1 [cf. (3.28)] 49
		3.3.2 An Exact Equilibration Algorithm for a Specially Structured Generalized Network 51
	3.4	A Case Study 51
	3.5	Summary and Conclusions 59
	3.6	Sources and Notes 60
	References 61

Part III Supply Chain Network Game Theory Models and Applications

4	**Food Supply Chains** 65
	4.1 Motivation and Overview 65
	4.2 The Food Supply Chain Network Oligopoly Model 66
	4.2.1 The Underlying Chemistry 69
	4.2.2 Competitive Behavior and Cournot–Nash Equilibrium 71
	4.3 The Algorithm 75
	4.3.1 Explicit Formulae for the Euler Method Applied to the Fresh Produce Supply Chain Network Oligopoly Variational Inequality (4.17) 76
	4.4 A Case Study 76
	4.5 Summary and Conclusions 83
	4.6 Sources and Notes 84
	References 85

5	**Pharmaceutical Supply Chains** 89
	5.1 Motivation and Overview 89
	5.2 The Pharmaceutical Supply Chain Network Oligopoly Model 91
	5.2.1 An Illustrative Pharmaceutical Supply Chain Network Example 96
	5.3 The Algorithm 102
	5.3.1 Explicit Formulae for the Euler Method Applied to the Pharmaceutical Supply Chain Generalized Network Oligopoly Variational Inequality (5.4) 102
	5.4 A Case Study 102
	5.5 Outsourcing 111
	5.6 Mergers and Acquisitions 112
	5.7 Summary and Conclusions 112
	5.8 Sources and Notes 113
	References 114

6	**Fast Fashion Apparel Supply Chains** 117
	6.1 Motivation and Overview 117
	6.2 The Sustainable Fashion Supply Chain Network Oligopoly Model 120

6.3	The Algorithm		127
	6.3.1	Explicit Formulae for the Euler Method Applied to the Sustainable Fashion Supply Chain Network Oligopoly Variational Inequality (6.14)	127
6.4	A Case Study		127
6.5	Summary and Conclusions		134
6.6	Sources and Notes		135
	References		137

Glossary of Notation 141

Part I
Introduction and Scope

Chapter 1
Introduction and Scope

Abstract In this chapter, we highlight the background behind and the motivation for a unified approach to supply chain network analytics for time-sensitive, perishable products. We describe the methodological framework and provide the organization of the book, detailing the featured applications in sectors ranging from healthcare, including pharmaceutical, blood, and medical nuclear supply chains, to food and fashion apparel that are explored in depth through models and case studies.

1.1 Background

Time is forever moving forward, weaving our economic activities into multidimensional supply chain networks across space. Products assembled from components and parts sourced from locations around the globe are transported to consumers thousands of miles away. Fruits and vegetables daily crisscross the equator to satisfy demanding consumers. Flowers are shipped across hemispheres to mark notable occasions. Radioisotopes and pharmaceuticals are processed and delivered across national boundaries to allow physicians to perform medical diagnostics and treatment in a timely manner. Fashion designers and apparel manufacturers respond to customers' evolving tastes by creating new styles within weeks. High-technology products from computers to consumer electronics decrease in value and functionality and are replaced as newer technologies and brands become available and customers clamor for the latest.

Today, the nature of production, transportation, and consumption of numerous products, as well as services, is global. Paradoxically, many products and, hence, their supply chains are increasingly time-sensitive—from medicines that must be consumed on schedule and in good condition in order that healing and recovery take place to fresh produce, fish, meat, and milk, whose quality deteriorates over time, and to healthcare products that require special storage conditions and may even decay over time.

We are now in an era of *Networks Against Time*, in which decision-makers, be they individuals, firms, or other organizations, who optimize and compete with and against time, not only will have the full advantages of the best allocation of their constrained resources but will also achieve their desired objectives.

For example, timely deliveries are becoming a strategy as important as productivity and even innovation—especially when the products are perishable. The time-sensitivity of a product may come from the characteristics of the product itself (as in the case of food) and/or be consumer-driven (as in the case of the demand for a particular consumer electronics product or fast fashion apparel). As Benjamin Franklin wrote in 1748 in his "Advice to a Young Tradesman, Written by an Old One," *Remember that TIME is Money* (cf. Labaree 1961). It may also be said that *time is life*, since time-sensitive products, such as vaccines and medicines, as well as, at the most fundamental level, food and water, are of a life-sustaining, if not, lifesaving, nature.

Time manifests itself in various forms in supply chain networks. Such activities as manufacturing, transportation, storage, and distribution to retailers and consumers have time directly associated with them. Time is also associated with the dynamics of adjustment processes. Firms must learn about their competitors' behavior as well as the consumers' preferences. Clearly, consumers are also time-sensitive when it comes to their purchases. Healthcare professionals require that medicines, vaccines, and radioisotopes be available when needed for treatments and diagnostics. Firms that delay production and delivery of products may lose not only sales but their reputation. Hence, the inclusion of time elements into supply chain analytics is critical, especially in today's global scenario in which not only are decision-makers personally pressed for time but the same holds for their organizational and business processes and products.

Today's supply chains are dynamic and complex, consisting of networks of manufacturers and suppliers, distributors, and consumers at the demand markets (Nagurney 2006). As the essential infrastructure for the production, distribution, and consumption of goods as well as services in today's globalized network economy, supply chains are widely exposed to ever-increasing challenges. Consumers' demand for new products, the deteriorating economic landscape, as well as environmental concerns require that companies be more creative while also becoming more rigorous and cost-effective in the procurement, production, and distribution of their products and services.

Nevertheless, despite significant achievements, the discipline of supply chain management is still incapable of satisfactorily handling many practical, real-world challenges (Georgiadis et al. 2005). When it comes to the supply chains of *perishable* products, that is, those with limited lifespans during which they should be consumed (Federgruen et al. 1986), these challenges are even of greater concern for decision-makers and managers.

We believe that a unified supply chain network analytics framework is needed. Such a framework should be able to handle optimization and competitive behavior and be relevant to the many industrial sectors in which perishable products are prominent, from healthcare to food to fast fashion apparel to high technology. With this volume, we hope to have made a contribution that is also conducive to further advances in both research and practice.

We note that classical examples of perishable goods include fresh produce in the form of, the already noted, fruits and vegetables, dairy products, medicines and vaccines, radioisotopes, cut flowers, and even human blood (cf. Ghare and Schrader 1963; Emmons 1968; Cohen and Pierskalla 1975; Aronson 1989; Nagurney and Aronson 1989; Raafat 1991; Ahumada and Villalobos 2009; Nahmias 2011; Nagurney and Nagurney 2012). However, in this book, we take the broader perspective of products being perishable not only in terms of their characteristics (such as their chemistry and the underlying physics) and *supply* (i.e., the manner of procurement/production/processing, storage, transportation, etc.) aspects but also in terms of the *demand* for the products. Hence, we also include under the *perishable product* umbrella products that are *discarded* (or replaced) relatively quickly after purchase, because of changing consumer tastes, such as *fast* fashion apparel, or those that become obsolete (as in certain high-technology products). Such an approach follows from Whitin (1957), who considered the deterioration of fashion goods at the end of a prescribed shortage period.

The supply chain management of perishable products at the strategic, tactical, and operational levels of the organizational decision-making hierarchy is faced with such challenges as:

- *Transportation and storage*: Many perishable products require careful handling, special transportation equipment, and cold storage facilities to ensure that the quality (and quantity) are preserved.
- *Inventory management*: At the demand points as well as at the other facilities of the supply chain network, inventory tracking and replenishment techniques may need to be utilized to minimize the outdating/deterioration of such products.
- *Incurred waste discarding cost*: Discarding of the waste associated with perished goods may impose additional costs to the firms.
- *Safety and environmental impact*: Perished products and the associated waste may be hazardous and may pollute the soil, the air, and the water.
- *Demand management*: Depending on the product and supply chain, demand may be uncertain or it may be known (as in scheduled treatments) and fixed. Demand may also be price-sensitive (as in fashion apparel, specific consumer goods, and certain pharmaceuticals).

1.2 The Methodological Framework

Unification is critical in scientific domains, from both theoretical and applied perspectives. It allows for consistency in terms of conceptualization, modeling, analysis, and applications. Moreover, a unifying framework can lead to additional advances and generate new insights since a powerful baseline that has been built can be further expanded upon.

In this book, since the applications of concern—those of perishable products—arise in numerous sectors of the economy, the methodological framework must be sufficiently general to capture not only the behavior of the decision-makers but also the scope and complexity of the underlying supply chains. It is essential that the perishability of the products over space and time also be appropriately formulated.

Since we are dealing with supply chains, a network formalism is mandated. It allows for a graphical depiction of the various facilities and the economic activities and processes, along with their relationships and interconnections. In addition, it is critical to capture rigorously the underlying decision-making behavior, whether that of a single decision-maker (manager/firm/organization) with responsibility for the entire supply chain network or the case of multiple competing decision-makers. In both cases, one needs to identify the appropriate objective functions (what one wants to achieve), what the variables that can be controlled and varied are, and what the constraints associated with them are.

Hence, the methodological framework for supply chains for perishable products must be able to handle different behavioral concepts, including optimization, as well as competitive equilibrium, and be sufficiently general to represent different supply chain structures that are relevant to the spectrum of perishable products.

We use, as the methodological framework, an integrated variational inequality theory approach, which combines network theory, with a focus on generalized networks, optimization, and game theory. Variational inequality theory allows for the modeling, analysis, and computation of solutions to problems under different behavioral concepts (e.g., optimization or Nash equilibrium) and enables the application of effective algorithms for computations, which, in the case of supply chain networks, can exploit the network structure. In addition, since solutions to variational inequality problems, under appropriate assumptions, also coincide with the solutions to associated projected dynamical systems, one can study how a supply chain network system evolves over time.

By utilizing a *generalized* network concept, one can track the perishability of products over space and time through the use of arc/link multipliers. Generalized networks contain, as a special case, *pure* networks if all the arc multipliers are set equal to one. In pure networks, the amount of flow (e.g., as of a product) that originates at a node and flows on an arc or link of a network is the same amount that arrives at the node that the arc terminates in. If an arc multiplier is less than one, then it is a *loss* multiplier; if it is greater than one, then it is a *gain* multiplier. Generalized networks have been used in applications ranging from agricultural products to finance (cf. Aronson 1989; Nagurney and Aronson 1989; Nagurney and Siokos 1997). Our first study on blood supply chain networks (Nagurney et al. 2012), as generalized networks, led us to additional research, which we integrate and synthesize in this book.

This book is meant to be self-contained, but we assume that the reader has sufficient knowledge of basic network theory and optimization theory. The supply chain network models described are all nonlinear, and the network structures are depicted graphically. Additional background material on variational inequality theory, projected dynamical systems, and dynamic supply chain networks can be found, respectively, in the books by Nagurney (1999), Nagurney and Zhang (1996), and Nagurney (2006). Geunes and Pardalos (2003) provided an annotated bibliography of network optimization to supply chain management. This is the first book to focus on supply chain network analytics for perishable products utilizing an integrated methodological framework that can handle either single or multiple decision-makers, coupled with the architecture of the supply chains. Analytics, the scientific process of transforming data into insight for making better decisions, provides the platform for the prescriptive decision-making.

1.3 The Organization of the Book

This book is comprised of three parts, beginning with this chapter, which makes up Part I and provides the introduction and the scope. Part II consists of two chapters, which make up the optimization component. Part III consists of three chapters, each of which contains a game theoretical supply chain network model. Relevant qualitative properties are given for each of the models, as well as algorithms, followed by a realistic case study. The input and output data are presented for completeness and transparency purposes. Each chapter includes a Sources and Notes section and a set of references.

Chapter 2 develops a supply chain network optimization model for blood banking. Blood is a highly perishable product and has motivated the development of the fundamentals of our framework. Chapter 3 constructs a medical nuclear supply chain optimization model for which the decay of the radioisotope is captured through arc multipliers calculated using physics principles. The models in Part II provide explicit multicriteria objective functions, since, as noted, perishable products also generate waste, with accompanying costs and risks.

Part III turns to supply chain network models and applications in which there are multiple decision-makers, specifically, firms, who are engaged in competition. The novelty of the models in Part III lies in that the firms produce substitutable products that are differentiated by *brand*. Hence, we utilize game theory.

Chapter 4 describes an application that anyone can relate to—that of food—specifically, fresh produce. In the model, we demonstrate how the arc multipliers can be obtained using the chemistry behind food deterioration. The firms seek to maximize their profits until an equilibrium is achieved. Chapter 5 returns to the healthcare stream of applications of Part II, but now the firms are pharmaceutical firms, who are profit maximizers. We investigate the impact of a new firm entering the market, after a drug patent expires.

Chapter 6 focuses on a fast fashion apparel application with brand differentiation. The model is a pure network model since the arc multipliers are all equal to one. As in the chapters in Part II, we utilize multicriteria decision-making, but rather than considering waste discarding costs and risk minimization, we address the emissions generated. Hence, an underlying theme of this book is *sustainability*. We also discuss how the model can be used for high-tech product supply chains.

References

Ahumada O, Villalobos JR (2009) Application of planning models in the agri-food supply chain: a review. Eur J Oper Res 196(1):1–20

Aronson JE (1989) A survey of dynamic network flows. Ann Oper Res 20:1–66

Cohen MA, Pierskalla WP (1975) Management policies for a regional blood bank. Transfusion 15(1):58–67

Emmons H (1968) A replenishment model for radioactive nuclide generators. Manag Sci 14(5):263–274

Federgruen A, Prastacos G, Zipkin PH (1986) An allocation and distribution model for perishable products. Oper Res 34(1):75–82

Georgiadis P, Vlachos D, Iakovou E (2005) A system dynamics modeling framework for the strategic supply chain management of food chains. J Food Eng 70(3):351–364

Geunes J, Pardalos PM (2003) Network optimization in supply chain management and financial engineering: An annotated bibliography. Networks 42(2):66–84

Ghare PM, Schrader GF (1963) A model for an exponentially decaying inventory. J Ind Eng XIV:238–243

Labaree LW (ed) (1961) The papers of Benjamin Franklin, vol 3, January 1, 1745 through June 30, 1750. Yale University Press, New Haven, Connecticut

Nagurney A (1999) Network economics: a variational inequality approach, 2nd and revised edn. Kluwer Academic, Dordrecht

Nagurney A (2006) Supply chain network economics: dynamics of prices, flows, and profits. Edward Elgar Publishing, Cheltenham

Nagurney A, Aronson J (1989) A general dynamic spatial price network equilibrium model with gains and losses. Networks 19(7):751–769

Nagurney A, Masoumi AH, Yu M (2012) Supply chain network operations management of a blood banking system with cost and risk minimization. Comput Manag Sci 9(2):205–231

Nagurney A, Nagurney LS (2012) Medical nuclear supply chain design: A tractable network model and computational approach. Int J Prod Econ 140(2):865–874

Nagurney A, Siokos S (1997) Financial networks: statics and dynamics. Springer, Heidelberg

Nagurney A, Zhang D (1996) Projected dynamical systems and variational inequalities with applications. Kluwer Academic, Boston, MA

Nahmias S (2011) Perishable inventory systems. Springer, New York

Raafat F (1991) Survey of literature on continuously deteriorating inventory models. J Oper Res Soc 42(1):27–37

Whitin TM (1957) The theory of inventory management, 2nd edn. Princeton University Press, Princeton, NJ

Part II
Supply Chain Network Optimization Models and Applications

Chapter 2
Blood Supply Chains

Abstract In this chapter, we develop a blood supply chain generalized network optimization model. The model captures the perishability of this lifesaving product through the use of arc multipliers. It contains discarding costs associated with waste/disposal, handles uncertainty associated with demand points, assesses costs associated with shortages/surpluses at the demand points, and quantifies the supply-side risk associated with procurement and other supply chain activities. A case study illustrates the applicability of the analytics framework.

2.1 Motivation and Overview

In this chapter, we focus on the optimization of a blood banking network system. Nahmias (1982) claimed that *The interest among researchers in perishable inventory problems has been sparked primarily by problems of blood bank management. Some of the possible reasons for the interest might be that blood bank research has been supported by public funds* (*thus providing incentives for academics*) *and hospitals and blood banks are not profit-making centers* (*thus making operational data more accessible*). This topic is especially timely today, since the number of disasters and the number of people affected by disasters have been growing over the past decade and blood is certainly a lifesaving product (cf. Nagurney and Qiang 2009). Because the shelf life of platelets is only 5 days and that of red blood cells 42 days, the blood supply chain is critically time-sensitive.

The availability of human blood is a key requirement of the worldwide healthcare system. According to the American Red Cross, the United States needs over 39,000 donations daily, and the blood supply is frequently reported to be just 2 days away from running out. Of 1,700 hospitals participating in a survey in 2007, over 25% reported cancellations of elective surgeries due to blood shortages. While for many hospitals the number of days of blood-related shortages was insignificant, some hospitals have reported delaying elective surgery on 50 or more days a year due to blood shortages. In 2012, the US blood supply reached its lowest level in 15 years (DiBlasio 2012). Furthermore, it was estimated in 2006 that of the 15.7 million units of blood and blood products in hospitals, almost 1.3 million units were beyond their expiration dates (Whitaker et al. 2007). In addition, the cost of a unit of red blood cells has increased 6.4% from 2005 to 2007, giving financial incentives for the efficient use of the blood supply. Moreover, hospitals are responsible for disposing approximately 90% of the outdated units of blood. The volume of this medical waste imposes a significant discarding cost to the already financially stressed hospitals (Chen 2010).

In this chapter, we develop a multicriteria *system-optimization* framework for a regionalized blood supply chain network. A system-optimization approach is mandated for critical supplies (Nagurney et al. 2011) in that the demand for such products must be satisfied as closely as possible at minimal total cost. The use of a profit maximization criterion is inappropriate, since nonprofit organizations with special status, such as the Red Cross, supply a large portion of the blood components.

Unlike many other supply chain models that assume a fixed lifetime for a perishable good (see, e.g., Hwang and Hahn 2000 and Zhou and Yang 2003), our system-optimization approach for supply chain network management captures the perishability/waste that occurs over the relevant links associated with various activities of the supply chain. Here we also take into account the discarding costs on the links as well as the discarding costs of outdated product at the demand points due to the possibility that excess supply was delivered. Furthermore, we capture in the model the uncertainty of the demand and the associated shortage penalties at the demand points.

This chapter is organized as follows. In Sect. 2.2, we describe, in detail, the structure of a regionalized blood banking system. We develop the blood supply chain network model and establish that the multicriteria optimization problem is equivalent to a variational inequality problem, with nice features for computations. We also present illustrative numerical examples and conduct sensitivity analysis. This model has several novel features:

- It captures perishability of this product through the use of arc multipliers.
- It contains discarding costs associated with waste/disposal.
- It handles uncertainty associated with demand points.
- It assesses costs associated with shortages/surpluses at the demand points.
- It quantifies the supply-side risk associated with procurement and other supply chain activities.

In Sect. 2.3, we propose an algorithm which, when applied, yields the optimal level of blood product flows. We then apply the algorithm to compute the solution to a case study in Sect. 2.4 using data motivated by a real-world application in order to further illustrate the modeling and computational framework. In Sect. 2.5, we summarize the results and present our conclusions. Section 2.6 contains the Sources and Notes for this chapter.

2.2 The Blood Supply Chain Network Model

In this section, we present the supply chain network model for regionalized blood banks. Although the structure of the blood supply chain network will differ among countries and regions, this network framework is sufficiently general to address any blood supply chain network.

2.2.1 The Components of a Blood Banking System

In many parts of the world, blood banking operations are conducted in a regionalized manner. There exists a *regional blood center* in each geographic area that is in charge of the coordination and administration of its lower-level units. In the USA, for example, the American Red Cross (ARC) oversees all of its regional blood banking operations. Other suppliers of blood include hospitals, which account for less than 5% of the market share (Whitaker et al. 2007) and private blood suppliers.

The American Red Cross has not charged for the blood since 1960. It only recovers the costs it incurs in the recruitment and screening of potential donors, the collection of blood by trained staff, the processing and testing of each unit of blood, and the labeling, storage, and distribution of blood components.

Figure 2.1 depicts the network topology of a regional blood banking system in the USA. In this network, the top level (origin) node represents the regional division. Every other node in the network denotes a component/facility in the system. A path connecting the origin node to a destination node (demand point) consists of a sequence of directed links which correspond to supply chain network activities that ensure that the blood is collected, processed, and, ultimately, distributed to the demand point. We assume that in the supply chain network topology, there exists at least one path joining node 1 with each destination node. This assumption guarantees that the demand at each demand point will be met as closely as possible, given that we will be considering uncertain demand for blood. The solution of this model yields the optimal flows of blood at minimum total cost and risk.

In the topology in Fig. 2.1, the first set of the links connecting the origin node to the second tier corresponds to the process of *blood collection*. The blood collection sites are denoted by $CS_1, CS_2, \ldots, CS_{n_{CS}}$, where n_{CS} is the number of such sites, constituting the second tier of the network. Many of these collection sites are mobile or temporary locations, while others are permanent sites. In the case of drastic shortages, due to natural or man-made disasters, for example, the regional divisions are likely to need to import blood products from other regions or even other countries, an aspect that is excluded from this model.

The third tier of nodes are the *blood centers*. There exist n_{BC} of these facilities in a region, denoted by $BC_1, BC_2, \ldots, BC_{n_{BC}}$. The whole blood is shipped to these facilities after being collected at the collection sites. The set of links connecting the second and third tiers of the network topology represents the *shipment of collected blood*.

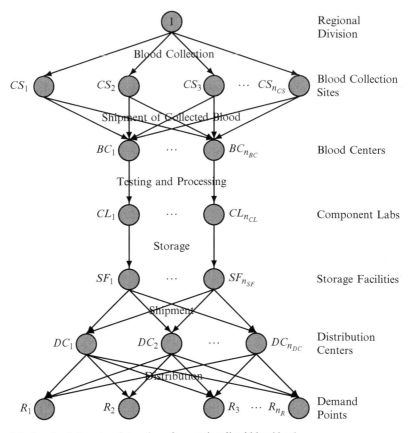

Fig. 2.1 Supply chain network topology for a regionalized blood bank

The fourth tier of the network is the processing facilities, commonly referred to as *component labs*. The number of these facilities in one region is assumed to be n_{CL} with the component labs denoted by $CL_1, \ldots, CL_{n_{CL}}$. They are, typically, located at the blood centers. At these labs, the collected blood is separated into parts, that is, red blood cells and plasma, since most recipients need only a specific component for transfusions. Every unit of donated whole blood can provide one unit of red blood cells and one unit of plasma. On the first three sets of links in Fig. 2.1, the flow is the amount of whole blood. The flow on the links thereafter denotes the number of units of red blood cells.

Plasma and other side derivatives are excluded from our model for several reasons. First, although plasma can be derived from donated whole blood, in practice, plasma is mainly produced in a different process called *apheresis*. Apheresis is a blood donation method where the blood is passed through an apparatus that separates out the plasma and returns the remainder, the red blood cells, to the donor. In addition, plasma can be stored frozen for an extended period of time, typically 1 year, which is not comparable to the approximately 6 week lifetime of red blood cells. Most importantly, whole blood and red blood cells account for the major part of donations and transfusions rather than plasma and other components (Whitaker et al. 2007).

The safety of the blood supply is considered to be the primary requirement for the blood banking system. US federal law mandates that every individual unit of donated blood be tested before being transfused, regardless of the number of the times a donor has previously donated blood. Blood is tested for multiple infectious disease markers, including HIV, hepatitis, and the West Nile virus (Redcrossblood.org 2010). The testing facilities are owned and operated by the American Red Cross and require heavy investments for specialized equipment. Presently, there are only five testing labs in the USA, which are shared among the 36 blood regions. A small sample of every donated blood unit is sent to the testing labs. These samples are discarded regardless of the results of the tests. Due to the high perishability of many of the blood products, the processes of testing and separating take place concurrently yet sometimes hundreds of miles away. If the result of a test on a unit of donated blood turns out to be positive, that unit will be discarded at the corresponding storage facility. In our model, the set of the links connecting the component labs to storage facilities corresponds to this *testing and processing*, and the costs on these links represent the operational cost of testing and processing combined. The fraction of the flow lost during or as a result of the testing process is also included in our model.

The fifth set of nodes in Fig. 2.1 denotes the short-term *storage facilities*. There are n_{SF} of such nodes in the network, denoted by $SF_1, SF_2, \ldots, SF_{n_{SF}}$, which are usually colocated with the component labs. The links connecting the component lab nodes to the storage facilities represent the *storage* activity of the tested and processed blood before it is shipped for distribution.

The next set of nodes in the supply chain network represents the *distribution centers*, denoted by $DC_1, DC_2, \ldots, DC_{n_{DC}}$, where n_{DC} is the total number of such facilities in the region. Distribution centers act as transshipment nodes and facilitate the distribution of blood to the destinations. The links connecting the storage tier to the distribution centers are *shipment* links.

The last set of links joining the bottom two tiers of the network in Fig. 2.1 are *distribution* links ending in n_R *demand points*. Hospitals and surgical medical centers are the predominant users of blood.

Specific elements of the supply chain network in Fig. 2.1 may be physically colocated with some others. Our supply chain network topology is process-based, rather than location-based, which is consistent with blood banking.

The supply chain network topology in Fig. 2.1 is denoted by $\mathscr{G} = [N, L]$, where N and L denote the sets of nodes and links, respectively.

2.2.2 The Formulation

The blood supply chain problem is formulated as a single-period type, where the time horizon spans the various activities of procurement, processing, and distribution.

Since whole blood is perishable, long-term storage is avoided, except for plasma, which is excluded from our model. Hence, the assumption of a single-period time horizon is realistic. The focus of this chapter is the operations management of the blood supply chain rather than its inventory management. Nevertheless, our model takes into account the potential shortage associated with the uncertain demand at the demand points. In addition, the surplus penalty addresses additional costs, due to excess supply or short-term inventory holding costs.

Associated with each link of the supply chain network is a unit operational cost function that reflects the cost of operating that particular supply chain activity. We denote the links by a, b, etc. The unit operational cost on link a is denoted by c_a and is a function of flow on that link, f_a. The *total* operational cost on link a is denoted by $\hat{c}_a$ and is constructed as

$$\hat{c}_a(f_a) = f_a \times c_a(f_a), \qquad \forall a \in L, \tag{2.1}$$

that is, the total cost on a link is equal to the unit cost on that link times the flow on that link. The link total cost functions are assumed to be convex and continuously differentiable.

The origin/destination (O/D) nodes consist of the pairs of nodes: $w_k \equiv (1, R_k)$; $k = 1, \ldots, n_R$, where $\mathscr{P}_k$ denotes the set of paths, which represent alternative associated series of supply chain network processes, joining $(1, R_k)$. $\mathscr{P}$ denotes the set of all paths joining node 1 to the destination nodes, and $n_{\mathscr{P}}$ denotes the number of paths.

Let v_k denote the *projected demand* for blood at the demand point R_k; $k = 1, \ldots, n_R$. We assume that the demand at each demand point is uncertain with a known probability distribution. The actual demand at demand point R_k; $k = 1, \ldots, n_R$, is denoted by d_k and is a random variable with probability density function given by $\mathscr{F}_k(t)$. Let P_k be the probability distribution function of d_k, that is, $P_k(D_k) = \text{Prob}(d_k \leq D_k) = \int_0^{D_k} \mathscr{F}_k(t) dt$.

Let Δ_k^- and Δ_k^+ represent the shortage and surplus of blood at demand point R_k, defined, respectively, by

$$\Delta_k^- \equiv \max\{0, d_k - v_k\}, \qquad k = 1, \ldots, n_R, \tag{2.2}$$

$$\Delta_k^+ \equiv \max\{0, v_k - d_k\}, \qquad k = 1, \ldots, n_R. \tag{2.3}$$

The expected values of the shortage (Δ_k^-) and the surplus (Δ_k^+) are given by

$$E(\Delta_k^-) = \int_{v_k}^{\infty} (t - v_k) \mathscr{F}_k(t) dt, \qquad k = 1, \ldots, n_R, \tag{2.4}$$

$$E(\Delta_k^+) = \int_0^{v_k} (v_k - t) \mathscr{F}_k(t) dt, \qquad k = 1, \ldots, n_R. \tag{2.5}$$

Due to the importance of the availability of blood at the demand points, a relatively large penalty of λ_k^- is associated with the shortage of a unit of blood at demand point R_k. Here, λ_k^- can include the social cost of a death or a severe

2.2 The Blood Supply Chain Network Model

injury of a patient, due to a blood shortage. Also, since blood is perishable and will become outdated if not used within a certain period after being delivered, an outdating penalty, λ_k^+, is assigned to a surplus unit of blood. Note that, in our formulation, this surplus penalty is charged to the organization even though it is not directly responsible for the outdated blood at the hospitals once it delivers it to them. This is because human blood is scarce, and the organization aims to minimize the amount of outdated blood at demand points. The disposal of outdated blood dominates the amount of medical waste generated during blood banking activities (Rios 2010). Hence, λ_k^+, in the case of blood (as for other perishable products), includes the cost of short-term inventory holding (cold storage) and, possibly, the discarding cost of the outdated product. Having outdated supplies at the demand points not only imposes a discarding cost on the already financially stressed healthcare institutions but also leads to further environmental damage. Similar examples of penalty costs, due to surpluses or shortages, can be found in the literature (see, e.g., Dong et al. 2004 and Nagurney et al. 2011). These penalties can be assessed by the authority who is contracting with the organization to deliver the blood.

Thus, the expected total penalty at demand point $R_k; k = 1, \ldots, n_R$, is

$$E(\lambda_k^- \Delta_k^- + \lambda_k^+ \Delta_k^+) = \lambda_k^- E(\Delta_k^-) + \lambda_k^+ E(\Delta_k^+). \tag{2.6}$$

The demand points are not the only elements of the blood supply chain in which the perishability of the collected blood occurs. Throughout the processes of blood collection, shipment, testing and processing, storage, and distribution, a fraction of the collected blood may deteriorate, be lost, or be wasted. This fraction depends on the activity since various processes lead to different amounts of waste and can also differ among the various facilities at the same tier of the network.

We associate with every link a in the supply chain network in Fig. 2.1 an arc/link multiplier α_a, which lies in the range of $(0,1]$. When the value of the multiplier α_a is 1, 100% of the initial flow on link a reaches its successor node, reflecting that there is no waste/loss on link a. The average percentage of loss due to the testing and processing was reported to be 1.7% (Sullivan et al. 2007). Therefore, the corresponding multiplier, α_a, for the testing and processing link would be equal to $1 - 0.017 = 0.983$.

Let f_a' denote the final flow on link a with initial flow f_a, that is, that flow that reaches the successor node of the link. Therefore,

$$f_a' = \alpha_a f_a, \quad \forall a \in L. \tag{2.7}$$

Thus, the waste/loss on link a is equal to

$$f_a - f_a' = (1 - \alpha_a) f_a, \quad \forall a \in L. \tag{2.8}$$

The organization is responsible for discarding this waste which is potentially regulated medical waste. Medical waste disposal contractors are typically employed to remove and dispose of the waste. Since α_a is assumed to be constant, and can be determined a priori, a discarding cost function, $\hat{z}_a$, which is a function of the flow,

f_a, and is assumed to be convex and continuously differentiable, is associated with each link and is of the form

$$\hat{z}_a = \hat{z}_a(f_a), \quad \forall a \in L. \tag{2.9}$$

Let x_p represent the (initial) flow of blood on path p joining the origin node with a destination node. All path flows must be nonnegative, that is,

$$x_p \geq 0, \quad \forall p \in \mathscr{P}, \tag{2.10}$$

since the blood is collected, tested, shipped, etc.

Let μ_p denote the path multiplier corresponding to the throughput on path p, which is defined as the product of all link multipliers on links comprising that path, that is,

$$\mu_p \equiv \prod_{a \in p} \alpha_a, \quad \forall p \in \mathscr{P}. \tag{2.11}$$

Although the amount of blood that originates on a path p is x_p, the amount (due to perishability) that actually arrives at the destination of this path is $x_p \mu_p$.

The path multipliers provide valuable information to decision-makers since they capture how much of the product that originates at an origin node of a path and *travels* along the path actually arrives at the destination node. If all the arc multipliers on a path p are positive but less than or equal to 1, then μ_p is also positive and less than or equal to 1. Hence, as a construct, μ_p is useful in modeling product perishability, as we will see in this and subsequent chapters.

The projected demand at demand point R_k, v_k, is the sum of all the final flows on paths joining $w_k = (1, R_k)$:

$$v_k = \sum_{p \in \mathscr{P}_k} x_p \mu_p, \quad k = 1, \ldots, n_R. \tag{2.12}$$

We define the arc–path multiplier, α_{ap}, which is the product of the multipliers of the links on path p that precede link a in that path, as

$$\alpha_{ap} \equiv \begin{cases} \delta_{ap} \prod_{\{a' < a\}} \alpha_{a'}, & \text{if } \{a' < a\} \neq \emptyset, \\ \delta_{ap}, & \text{if } \{a' < a\} = \emptyset, \end{cases} \tag{2.13}$$

where $\{a' < a\}$ denotes the set of the links preceding link a in path p, δ_{ap} is defined as equal to 1 if link a is contained in path p, and 0, otherwise, and $\emptyset$ denotes the null set. If link a is not contained in path p, then α_{ap} is set to zero. If a belongs to the first set of links, the blood collection links, this multiplier is equal to 1. The relationship between the link flow, f_a, and the path flows can, hence, be expressed as

$$f_a = \sum_{p \in \mathscr{P}} x_p \, \alpha_{ap}, \quad \forall a \in L. \tag{2.14}$$

2.2 The Blood Supply Chain Network Model

Similar examples of multipliers corresponding to the loss/waste on links or paths can be found in the literature (see, e.g., Nagurney and Aronson 1989). Of course, if all the arc multipliers on a path are greater than 1, that is, they are gain, rather than loss, multipliers, then the corresponding path multiplier will also be greater than 1. Gain arc and path multipliers can capture gains in financial applications as well as in agricultural ones.

We group the path flows into the vector x. Also, the link flows, and the projected demands are grouped into the respective vectors f and v.

The total cost minimization objective faced by the organization includes the total cost of operating the various links, the total discarding cost of waste/loss over the links, and the expected total blood supply shortage cost as well as the total discarding cost of outdated blood at the demand points. This optimization problem can be expressed as

$$\text{Minimize} \quad \sum_{a \in L} \hat{c}_a(f_a) + \sum_{a \in L} \hat{z}_a(f_a) + \sum_{k=1}^{n_R} \left(\lambda_k^- E(\Delta_k^-) + \lambda_k^+ E(\Delta_k^+) \right) \quad (2.15)$$

subject to: constraints (2.10), (2.12), and (2.14).

As mentioned earlier, the minimization of total costs is not the only objective of suppliers of perishable goods. One of the most significant challenges is to capture the risk associated with different activities in the blood supply chain network. Unlike the demand that can be projected according to the historical data, the amount of donated blood at the collection sites has been observed to be highly stochastic.

Interestingly, disasters, such as the 2010 earthquake in Haiti, may stimulate people's sympathy and dramatically increase the number of blood donors. As in Nagurney et al. (2005), we introduce a total risk function $\hat{r}_a$ corresponding to link a, $\forall a \in L$, and emphasize that, in the case of the procurement links, such a function is especially relevant. This function is assumed to be convex and continuously differentiable and a function of the flow on its link. The organization attempts to minimize the total risk over all links of the network.

Thus, the risk minimization objective function for the organization can be expressed as

$$\text{Minimize} \quad \sum_{a \in L} \hat{r}_a(f_a), \quad (2.16)$$

where $\hat{r}_a = \hat{r}_a(f_a)$ is the total risk function on link a. We associate with the total risk objective, (2.16), a risk-aversion factor θ, which is assigned by the decision-maker.

The blood supply chain network optimization problem can be expressed as the multicriteria decision-making problem (cf. Fishburn 1970; Chankong and Haimes 1983; Keeney and Raiffa 1993; Nagurney and Dong 2002):

$$\text{Minimize} \quad \sum_{a \in L} \hat{c}_a(f_a) + \sum_{a \in L} \hat{z}_a(f_a) + \sum_{k=1}^{n_R} \left(\lambda_k^- E(\Delta_k^-) + \lambda_k^+ E(\Delta_k^+) \right) + \theta \sum_{a \in L} \hat{r}_a(f_a) \quad (2.17)$$

subject to constraints (2.10), (2.12), and (2.14).

The above optimization problem is in terms of link flows. It can also be expressed (cf. (2.14)) in terms of path flows:

$$\text{Minimize} \sum_{p \in \mathscr{P}} (\hat{C}_p(x) + \hat{Z}_p(x)) + \sum_{k=1}^{n_R} (\lambda_k^- E(\Delta_k^-) + \lambda_k^+ E(\Delta_k^+)) + \theta \sum_{p \in \mathscr{P}} \hat{R}_p(x)$$
(2.18)

subject to constraints (2.10) and (2.12). The total operational cost, $\hat{C}_p(x)$; the total discarding cost, $\hat{Z}_p(x)$; and the total risk, $\hat{R}_p(x)$, corresponding to path p are, respectively, derived from $C_p(x)$, $Z_p(x)$, and $R_p(x)$ as follows:

$$\hat{C}_p(x) = x_p \times C_p(x), \quad \hat{Z}_p(x) = x_p \times Z_p(x), \quad \hat{R}_p(x) = x_p \times R_p(x), \quad \forall p \in \mathscr{P},$$
(2.19)

with the unit cost functions on path p, that is, $C_p(x), Z_p(x)$, and $R_p(x)$, in turn, as below:

$$C_p(x) \equiv \sum_{a \in L} c_a(f_a) \alpha_{ap}, \quad Z_p(x) \equiv \sum_{a \in L} z_a(f_a) \alpha_{ap}, \quad R_p(x) \equiv \sum_{a \in L} r_a(f_a) \alpha_{ap}, \forall p \in \mathscr{P}.$$
(2.20)

Next, we present some preliminaries that enable us to express the partial derivatives of the expected total shortage costs and the expected total surplus costs of outdated blood at the demand points solely in terms of path flow variables. Observe that, for each O/D pair $w_k = (1, R_k)$,

$$\frac{\partial E(\Delta_k^-)}{\partial x_p} = \frac{\partial E(\Delta_k^-)}{\partial v_k} \times \frac{\partial v_k}{\partial x_p}, \quad \forall p \in \mathscr{P}_k; k = 1, \ldots, n_R. \quad (2.21)$$

By Leibniz's integral rule, we have

$$\frac{\partial E(\Delta_k^-)}{\partial v_k} = \frac{\partial}{\partial v_k} \left(\int_{v_k}^{\infty} (t - v_k) \mathscr{F}_k(t) dt \right) = \int_{v_k}^{\infty} \frac{\partial}{\partial v_k} (t - v_k) \mathscr{F}_k(t) dt$$
$$= P_k(v_k) - 1, \quad k = 1, \ldots, n_R. \quad (2.22a)$$

Therefore,

$$\frac{\partial E(\Delta_k^-)}{\partial v_k} = P_k \left(\sum_{q \in \mathscr{P}_k} x_q \mu_q \right) - 1, \quad k = 1, \ldots, n_R. \quad (2.22b)$$

On the other hand, we have

$$\frac{\partial v_k}{\partial x_p} = \frac{\partial}{\partial x_p} \sum_{q \in \mathscr{P}_k} x_q \mu_q = \mu_p, \quad \forall p \in \mathscr{P}_k; k = 1, \ldots, n_R. \quad (2.23)$$

The above is correct since the μ_ps are constant values. Therefore, from (2.22b) and (2.23), we conclude that

2.2 The Blood Supply Chain Network Model

$$\frac{\partial E(\Delta_k^-)}{\partial x_p} = \mu_p \left[P_k \left(\sum_{q \in \mathscr{P}_k} x_q \mu_q \right) - 1 \right], \quad \forall p \in \mathscr{P}_k; k = 1, \ldots, n_R. \tag{2.24}$$

Similarly, for the surplus, we have

$$\frac{\partial E(\Delta_k^+)}{\partial x_p} = \frac{\partial E(\Delta_k^+)}{\partial v_k} \times \frac{\partial v_k}{\partial x_p}, \quad \forall p \in \mathscr{P}_k; k = 1, \ldots, n_R, \tag{2.25}$$

$$\frac{\partial E(\Delta_k^+)}{\partial v_k} = \frac{\partial}{\partial v_k} \left(\int_0^{v_k} (v_k - t) \mathscr{F}_k(t) dt \right) = \int_0^{v_k} \frac{\partial}{\partial v_k} (v_k - t) \mathscr{F}_k(t) dt$$
$$= P_k(v_k), \quad k = 1, \ldots, n_R. \tag{2.26a}$$

Thus,

$$\frac{\partial E(\Delta_k^+)}{\partial v_k} = P_k \left(\sum_{q \in \mathscr{P}_k} x_q \mu_q \right), \quad k = 1, \ldots, n_R. \tag{2.26b}$$

From (2.26b) and (2.23) we obtain

$$\frac{\partial E(\Delta_k^+)}{\partial x_p} = \mu_p P_k \left(\sum_{q \in \mathscr{P}_k} x_q \mu_q \right), \quad \forall p \in \mathscr{P}_k; k = 1, \ldots, n_R. \tag{2.27}$$

Let K denote the feasible set such that

$$K \equiv \{x | x \in R_+^{n_{\mathscr{P}}}\}. \tag{2.28}$$

Before deriving two variational inequality formulations of the problem, we establish a lemma that formalizes the construction of the partial derivatives of the path total operational cost, the total discarding cost, and the total risk with respect to a path flow.

Lemma 2.1. *The partial derivatives of the total operational cost, the total discarding cost, and the total risk with respect to the corresponding path flow are, respectively, given by*

$$\frac{\partial (\sum_{q \in \mathscr{P}} \hat{C}_q(x))}{\partial x_p} = \sum_{a \in L} \frac{\partial \hat{c}_a(f_a)}{\partial f_a} \alpha_{ap}, \quad \forall p \in \mathscr{P}, \tag{2.29a}$$

$$\frac{\partial (\sum_{q \in \mathscr{P}} \hat{Z}_q(x))}{\partial x_p} = \sum_{a \in L} \frac{\partial \hat{z}_a(f_a)}{\partial f_a} \alpha_{ap}, \quad \forall p \in \mathscr{P}, \tag{2.29b}$$

$$\frac{\partial (\sum_{q \in \mathscr{P}} \hat{R}_q(x))}{\partial x_p} = \sum_{a \in L} \frac{\partial \hat{r}_a(f_a)}{\partial f_a} \alpha_{ap}, \quad \forall p \in \mathscr{P}. \tag{2.29c}$$

Proof. We establish the equivalence for (2.29a); the equivalences for (2.29b) and (2.29c) can be obtained in a similar fashion. The partial derivative of the total operational cost with respect to the flow on path p is first defined as

$$\frac{\partial (\sum_{q \in \mathscr{P}} \hat{C}_q)}{\partial x_p} = \sum_{q \in \mathscr{P}} \frac{\partial \hat{C}_q}{\partial x_p}, \quad \forall p \in \mathscr{P}, \tag{2.30a}$$

which, based on the total path cost (2.19), can be rewritten as

$$\frac{\partial (\sum_{q \in \mathscr{P}} \hat{C}_q)}{\partial x_p} = \sum_{q \in \mathscr{P}} \frac{\partial (C_q x_q)}{\partial x_p} = C_p + \sum_{q \in \mathscr{P}} x_q \frac{\partial C_q}{\partial x_p}, \quad \forall p \in \mathscr{P}. \tag{2.30b}$$

According to $C_q(x)$ in (2.20), we have

$$\frac{\partial C_q}{\partial x_p} = \frac{\partial \sum_{a \in L} c_a \alpha_{aq}}{\partial x_p} = \sum_{a \in L} \frac{\partial c_a}{\partial x_p} \alpha_{aq} = \sum_{a \in L} \frac{\partial c_a}{\partial f_a} \frac{\partial f_a}{\partial x_p} \alpha_{aq}, \quad \forall p, q \in \mathscr{P}. \tag{2.31}$$

On the other hand, by referring to (2.14) yields

$$\frac{\partial f_a}{\partial x_p} = \alpha_{ap}, \quad \forall a \in L, \forall p \in \mathscr{P}. \tag{2.32}$$

From (2.31) and (2.32), we obtain

$$\frac{\partial C_q}{\partial x_p} = \sum_{a \in L} \frac{\partial c_a}{\partial f_a} \alpha_{ap} \alpha_{aq}, \quad \forall p, q \in \mathscr{P}. \tag{2.33}$$

Substituting (2.33) into (2.30b) yields

$$\frac{\partial (\sum_{q \in \mathscr{P}} \hat{C}_q)}{\partial x_p} = C_p + \sum_{q \in \mathscr{P}} x_q \sum_{a \in L} \frac{\partial c_a}{\partial f_a} \alpha_{ap} \alpha_{aq} = C_p + \sum_{a \in L} \sum_{q \in \mathscr{P}} x_q \frac{\partial c_a}{\partial f_a} \alpha_{ap} \alpha_{aq}$$

$$= C_p + \sum_{a \in L} \frac{\partial c_a}{\partial f_a} \alpha_{ap} \sum_{q \in \mathscr{P}} x_q \alpha_{aq}, \quad \forall p \in \mathscr{P}. \tag{2.34}$$

By substituting proper equivalences from (2.14) and (2.20) into (2.34), we have

$$\frac{\partial (\sum_{q \in \mathscr{P}} \hat{C}_q)}{\partial x_p} = \sum_{a \in L} c_a \alpha_{ap} + \sum_{a \in L} \frac{\partial c_a}{\partial f_a} \alpha_{ap} f_a = \sum_{a \in L} \left(c_a + \frac{\partial c_a}{\partial f_a} f_a \right) \alpha_{ap}, \quad \forall p \in \mathscr{P}. \tag{2.35}$$

On the other hand, from (2.1),

$$\frac{\partial \hat{c}_a}{\partial f_a} = c_a + \frac{\partial c_a}{\partial f_a} f_a, \quad \forall a \in L. \tag{2.36}$$

2.2 The Blood Supply Chain Network Model

From (2.35) and (2.36), we conclude that

$$\frac{\partial(\sum_{q\in\mathscr{P}}\hat{C}_q)}{\partial x_p} = \sum_{a\in L}\frac{\partial\hat{c}_a(f_a)}{\partial f_a}\alpha_{ap}, \quad \forall p \in \mathscr{P}. \tag{2.37}$$

Thus, (2.29a) has been established. □

We now derive the variational inequality formulations of the blood supply chain network optimization problem in terms of path flows and link flows.

Theorem 2.1 (Variational Inequality Formulations). *The vector x^* is an optimal solution to the multicriteria optimization problem (2.18), subject to (2.10) and (2.12), if and only if it is a solution to the variational inequality problem: determine the vector of optimal path flows $x^* \in K$, such that*

$$\sum_{k=1}^{n_R}\sum_{p\in\mathscr{P}_k}\left[\frac{\partial(\sum_{q\in\mathscr{P}}\hat{C}_q(x^*))}{\partial x_p} + \frac{\partial(\sum_{q\in\mathscr{P}}\hat{Z}_q(x^*))}{\partial x_p} + \lambda_k^+\mu_p P_k\left(\sum_{q\in\mathscr{P}_k}x_q^*\mu_q\right)\right.$$

$$\left.-\lambda_k^-\mu_p\left(1 - P_k\left(\sum_{q\in\mathscr{P}_k}x_q^*\mu_q\right)\right) + \theta\frac{\partial(\sum_{q\in\mathscr{P}}\hat{R}_q(x^*))}{\partial x_p}\right] \times [x_p - x_p^*] \geq 0, \quad \forall x \in K.$$

(2.38)

The variational inequality (2.38), in turn, can be rewritten in terms of link flows as follows: determine the vector of optimal link flows and the vector of optimal projected demands $(f^*, v^*) \in K^1$, such that

$$\sum_{a\in L}\left[\frac{\partial\hat{c}_a(f_a^*)}{\partial f_a} + \frac{\partial\hat{z}_a(f_a^*)}{\partial f_a} + \theta\frac{\partial\hat{r}_a(f_a^*)}{\partial f_a}\right] \times [f_a - f_a^*]$$

$$+ \sum_{k=1}^{n_R}\left[\lambda_k^+ P_k(v_k^*) - \lambda_k^-(1 - P_k(v_k^*))\right] \times [v_k - v_k^*] \geq 0, \quad \forall (f, v) \in K^1, \quad (2.39)$$

where K^1 denotes the feasible set as defined below:

$$K^1 \equiv \{(f, v) | \exists x \geq 0, \text{ and } (2.12) \text{ and } (2.14) \text{ hold}\}.$$

Proof. First, we prove the result for path flows [cf. (2.38)].

The convexity of $\hat{C}_p$, $\hat{Z}_p$, and $\theta\hat{R}_p$ for all paths p holds since $\hat{c}_a$, $\hat{z}_a$, and $\hat{r}_a$ were assumed to be convex for all links a and θ is nonnegative. Hence, their sum over all paths, as in (2.18), is also convex. We need to verify that $\lambda_k^- E(\Delta_k^-) + \lambda_k^+ E(\Delta_k^+)$ is also convex. We have that for all paths $p \in \mathscr{P}_k$; $k = 1, \ldots, n_R$:

$$\frac{\partial^2}{\partial x_p^2}\left[\lambda_k^- E(\Delta_k^-) + \lambda_k^+ E(\Delta_k^+)\right] = \lambda_k^-\frac{\partial^2 E(\Delta_k^-)}{\partial x_p^2} + \lambda_k^+\frac{\partial^2 E(\Delta_k^+)}{\partial x_p^2}. \tag{2.40a}$$

Substituting the first order derivatives from (2.24) and (2.27) into (2.40a) yields

$$\frac{\partial^2}{\partial x_p^2}\left[\lambda_k^- E(\Delta_k^-) + \lambda_k^+ E(\Delta_k^+)\right]$$

$$= \lambda_k^- \frac{\partial}{\partial x_p} \mu_p \left[P_k\left(\sum_{q\in\mathscr{P}_k} x_q\mu_q\right) - 1\right] + \lambda_k^+ \frac{\partial}{\partial x_p}\mu_p P_k\left(\sum_{q\in\mathscr{P}_k} x_q\mu_q\right)$$

$$= (\lambda_k^- + \lambda_k^+)(\mu_p)^2 \mathscr{F}_k\left(\sum_{q\in\mathscr{P}_k} x_q\mu_q\right) > 0, \quad \forall p \in \mathscr{P}_k; k=1,\ldots,n_R. \quad (2.40b)$$

The above inequality holds provided that $(\lambda_k^- + \lambda_k^+)$, that is, the sum of shortage and surplus penalties, is assumed to be positive. Hence, $\lambda_k^- E(\Delta_k^-) + \lambda_k^+ E(\Delta_k^+)$, and, as a consequence, the multicriteria objective function in (2.18) is also convex.

Since the objective function (2.18) is convex and the feasible set K is closed and convex, the variational inequality (2.38) follows from the standard theory of variational inequalities (cf. Nagurney 1999).

As for the proof of the variational inequality (2.39), now that (2.38) is established, we can apply Lemma 2.1. Also, from (2.12) and (2.14), we can rewrite the formulation in terms of link flows and projected demands rather than path flows. Thus, the second part of Theorem 2.1, that is, the variational inequality in link flows (2.39), holds. □

Variational inequality (2.38) can be put into standard form (see Nagurney 1999), the definition of which follows.

Definition 2.1 (Variational Inequality Problem). The finite-dimensional variational inequality problem VI$(F,\mathscr{K})$ is to determine the vector $X^* \in \mathscr{K} \subset R^n$, such that

$$\langle F(X^*), X - X^* \rangle \geq 0, \quad \forall X \in \mathscr{K}, \quad (2.41)$$

where F is a given continuous function from $\mathscr{K}$ to R^n and $\mathscr{K}$ is a given closed, convex set, with $\langle \cdot,\cdot \rangle$ denoting the inner product in n-dimensional Euclidean space.

Indeed, if we define the feasible set $\mathscr{K} \equiv K$ and the vectors $X \equiv x$ and $F(X)$, such that

$$F(X) \equiv \left[\frac{\partial \sum_{q\in\mathscr{P}} \hat{C}_q(x)}{\partial x_p} + \frac{\partial \sum_{q\in\mathscr{P}} \hat{Z}_q(x)}{\partial x_p} + \lambda_k^+ \mu_p P_k\left(\sum_{q\in\mathscr{P}_k} x_q\mu_q\right)\right.$$

$$\left.-\lambda_k^- \mu_p\left(1 - P_k\left(\sum_{q\in\mathscr{P}_k} x_q\mu_q\right)\right) + \theta \frac{\partial \sum_{q\in\mathscr{P}} \hat{R}_q(x)}{\partial x_p}; \quad p \in \mathscr{P}_k; k=1,\ldots,n_R\right], \quad (2.42)$$

then the variational inequality (2.38) can be reexpressed in standard form (2.41).

2.2 The Blood Supply Chain Network Model 25

We utilize variational inequality (2.38) in path flows for our computations since our proposed computational procedure will yield closed form expressions at each iteration. Once we have solved problem (2.38), by using (2.14), which relates the link flows to the path flows, we can obtain the solution f^* that minimizes the total cost as well as the total supply risk [cf. (2.17)] associated with the optimization of the supply chain network of a regionalized blood banking system.

2.2.3 Illustrative Blood Supply Chain Network Examples

We now present several numerical blood supply chain network examples for illustrative purposes.

The organization has a single blood collection site, a single blood center, one component lab, one storage facility, and a single distribution center and is to serve a single demand point. The links are labeled in Fig. 2.2 as a, b, c, d, e, and f.

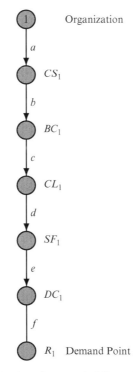

Fig. 2.2 Supply chain network topology for numerical Examples 2.1 and 2.2

Example 2.1. The total cost functions on the links were

$$\hat{c}_a(f_a) = f_a^2 + 6f_a, \; \hat{c}_b(f_b) = 2f_b^2 + 7f_b, \; \hat{c}_c(f_c) = f_c^2 + 11f_c, \; \hat{c}_d(f_d) = 3f_d^2 + 11f_d,$$
$$\hat{c}_e(f_e) = f_e^2 + 2f_e, \quad \hat{c}_f(f_f) = f_f^2 + f_f.$$

We assumed that there was no waste so that $\alpha_a = 1$ for all links in Fig. 2.2. Hence, all the functions $\hat{z}_a$ were set equal to 0 for all the links $a, \ldots, f$. The total risk cost function on the blood collection link a was $\hat{r}_a = 2f_a^2$, with the risk functions for all other links being set equal to zero. The risk-aversion factor, θ, was assumed to be 1.

There is only a single path p_1 which was defined as $p_1 = (a, b, c, d, e, f)$ with the path multiplier $\mu_{p_1} = 1$.

We assumed that the demand for the product followed a uniform distribution on the interval $[0, 5]$ so that $P_1(x_{p_1}) = \frac{x_{p_1}}{5}$. The penalties were $\lambda_1^- = 100$, $\lambda_1^+ = 0$.
Substitution of the values of $\lambda_1^+, \lambda_1^-, \mu_{p_1}$, and θ into (2.38) yields

$$\left[\frac{\partial \hat{C}_{p_1}(x^*)}{\partial x_{p_1}} - 100(1 - P_1(x_{p_1}^*)) + \frac{\partial \hat{R}_{p_1}(x^*)}{\partial x_{p_1}}\right] \times [x_{p_1} - x_{p_1}^*] \geq 0, \; \forall x \in K. \quad (2.43)$$

Under the assumption that $x_{p_1}^* > 0$, the left-hand side of inequality (2.43) must be equal to zero, that is,

$$\frac{\partial \hat{C}_{p_1}(x^*)}{\partial x_{p_1}} - 100(1 - P_1(x_{p_1}^*)) + \frac{\partial \hat{R}_{p_1}(x^*)}{\partial x_{p_1}} = 0. \quad (2.44)$$

It follows from Lemma 2.1 that

$$\frac{\partial \hat{C}_{p_1}(x^*)}{\partial x_{p_1}} = \frac{\partial \hat{c}_a(f_a^*)}{\partial f_a}\alpha_{ap_1} + \frac{\partial \hat{c}_b(f_b^*)}{\partial f_b}\alpha_{bp_1} + \frac{\partial \hat{c}_c(f_c^*)}{\partial f_c}\alpha_{cp_1} + \frac{\partial \hat{c}_d(f_d^*)}{\partial f_d}\alpha_{dp_1}$$
$$+ \frac{\partial \hat{c}_e(f_e^*)}{\partial f_e}\alpha_{ep_1} + \frac{\partial \hat{c}_f(f_f^*)}{\partial f_f}\alpha_{fp_1}. \quad (2.45)$$

Since the arc-path multipliers in this example are such that $\alpha_{ap_1} = \alpha_{bp_1} = \alpha_{cp_1} = \alpha_{dp_1} = \alpha_{ep_1} = \alpha_{fp_1} = 1$, it follows that the link flows satisfy $f_a^* = f_b^* = f_c^* = f_d^* = f_e^* = f_f^* = x_{p_1}^*$. Substitution of these equivalences into (2.45) gives us

$$\frac{\partial \hat{C}_{p_1}(x^*)}{\partial x_{p_1}} = (2f_a^* + 6) + (4f_b^* + 7) + (2f_c^* + 11) + (6f_d^* + 11) + (2f_e^* + 2)$$
$$+ (2f_f^* + 1) = 18x_{p_1}^* + 38. \quad (2.46)$$

Similarly,

$$\frac{\partial \hat{R}_{p_1}(x^*)}{\partial x_{p_1}} = \frac{\partial \hat{r}_a(f_a^*)}{\partial f_a}\alpha_{ap_1} = 4f_a^* = 4x_{p_1}^*. \quad (2.47)$$

Therefore, using the above relationships, (2.44) may be reexpressed as

$$18x_{p_1}^* + 38 - 100\left(1 - \frac{x_{p_1}^*}{5}\right) + 4x_{p_1}^* = 0, \quad (2.48)$$

2.2 The Blood Supply Chain Network Model

whose solution yields the optimal path flow, $x_{p_1}^* = 1.48$ and the corresponding optimal link flow pattern, $f_a^* = f_b^* = f_c^* = f_d^* = f_e^* = f_f^* = 1.48$. Following (2.12), the projected demand is equal to $v_1^* = x_{p_1}^* = 1.48$.

Example 2.2. Example 2.2 had the same data as Example 2.1 except that now there was a loss associated with the testing and processing link with $\alpha_c = .8$. Hence, we now set [cf. (2.9)] $\hat{z}_c = .5 f_c^2$ and $\mu_{p_1} = \alpha_c = .8$.

Similar to the solution procedure used for Example 2.1, from variational inequality formulation (2.38), under the assumption that $x_{p_1}^* > 0$, the following equation must hold for Example 2.2:

$$\frac{\partial \hat{C}_{p_1}(x^*)}{\partial x_{p_1}} + \frac{\partial \hat{Z}_{p_1}(x^*)}{\partial x_{p_1}} - 100 \times 0.8(1 - P_1(0.8 \times x_{p_1}^*)) + \frac{\partial \hat{R}_{p_1}(x^*)}{\partial x_{p_1}} = 0. \quad (2.49)$$

Since in this example, $\alpha_{ap_1} = \alpha_{bp_1} = \alpha_{cp_1} = 1$, $\alpha_{dp_1} = \alpha_{ep_1} = \alpha_{fp_1} = 0.8$, $f_a^* = f_b^* = f_c^* = x_{p_1}^*$, and $f_d^* = f_e^* = f_f^* = 0.8 x_{p_1}^*$, therefore,

$$\frac{\partial \hat{C}_{p_1}(x^*)}{\partial x_{p_1}} = (2 f_a^* + 6) + (4 f_b^* + 7) + (2 f_c^* + 11) + 0.8(6 f_d^* + 11) + 0.8(2 f_e^* + 2)$$

$$+ 0.8(2 f_f^* + 1) = 8 x_{p_1}^* + 24 + 0.8(10 \times 0.8 x_{p_1}^* + 14) = 14.4 x_{p_1}^* + 35.2. \quad (2.50)$$

Also,

$$\frac{\partial \hat{Z}_{p_1}(x^*)}{\partial x_{p_1}} = \frac{\partial \hat{z}_c(f_c^*)}{\partial f_c} \alpha_{cp_1} = f_c^* = x_{p_1}^*. \quad (2.51)$$

The partial derivative of the total risk function was equal to that of Example 2.1:

$$\frac{\partial \hat{R}_{p_1}(x^*)}{\partial x_{p_1}} = \frac{\partial \hat{r}_a(f_a^*)}{\partial f_a} \alpha_{ap_1} = 4 f_a^* = 4 x_{p_1}^*. \quad (2.52)$$

However, now we have

$$P_1(0.8 x_{p_1}^*) = \frac{0.8 x_{p_1}^*}{5}. \quad (2.53)$$

Therefore, the following equation needs to be solved:

$$14.4 x_{p_1}^* + 35.2 + x_{p_1}^* - 80 \left(1 - \frac{0.8 x_{p_1}^*}{5}\right) + 4 x_{p_1}^* = 0. \quad (2.54)$$

The new optimal path flow solution was $x_{p_1}^* = 1.39$, which corresponds to the optimal link flow pattern: $f_a^* = f_b^* = f_c^* = 1.39$ and $f_d^* = f_e^* = f_f^* = 1.11$. The projected demand was $v_1^* = x_{p_1}^* \mu_{p_1} = 1.11$. Comparing the results of Examples 2.1 and 2.2 reveals the fact that when perishability is taken into consideration, with $\alpha_c = .8$ and the above data, the organization chooses to produce/ship slightly smaller quantities so as to minimize the discarding cost of the waste, despite the shortage penalty of λ_1^-.

Note that when $\lambda_1^- = 200$, the optimal path flow solution becomes $x_{p_1}^* = 2.77$, and the corresponding optimal link flow pattern $f_a^* = f_b^* = f_c^* = 2.77$ and $f_d^* = f_e^* = f_f^* = 2.22$, with a projected demand of $v_1^* = x_{p_1}^* \mu_{p_1} = 2.22$.

By replacing the 100 in Eq. (2.49) with λ_1^-, we obtain the optimal path flow $x_{p_1}^*$ as a function of λ_1^-:

$$x_{p_1}^* = \frac{100(\lambda_1^- - 44)}{16\lambda_1^- + 2425}. \tag{2.55}$$

Therefore, an appropriate increase in the unit shortage penalty cost λ_1^- results in the organization processing larger volumes of the blood product, even exceeding the optimal flow in Example 2.1, which makes sense intuitively. Note that, for $\lambda_1^- \leq 44$, the organization should acquire and, hence, process and distribute zero units of the blood product.

2.2.3.1 Sensitivity Analysis

We conducted additional sensitivity analysis for Example 2.2 by varying the arc multiplier, α_c, and the unit shortage penalty cost, λ_1^-. The computed optimal path flow and the corresponding value of the objective function (2.18), denoted by OFV*, are reported in Table 2.1.

Table 2.1 Computed optimal path flows $x_{p_1}^*$ and values of the objective function at the optimal solution as α_c and λ_1^- vary

λ_1^-	α_c	.2	.4	.6	.8	1
100	$x_{p_1}^*$	0.00	0.58	1.16	1.39	1.48
	OFV*	250.00	246.96	234.00	218.83	204.24
200	$x_{p_1}^*$	0.88	2.40	2.83	2.77	2.61
	OFV*	494.19	439.52	376.23	326.94	288.35
300	$x_{p_1}^*$	2.10	3.74	3.86	3.54	3.20
	OFV*	715.12	581.15	464.85	387.17	331.44
400	$x_{p_1}^*$	3.20	4.76	4.57	4.03	3.55
	OFV*	914.75	689.71	525.36	425.56	357.63
500	$x_{p_1}^*$	4.20	5.57	5.09	4.37	3.79
	OFV*	1,096.03	775.55	569.30	452.16	375.23
1000	$x_{p_1}^*$	8.09	7.95	6.41	5.19	4.33
	OFV*	1,799.11	1,027.94	681.89	515.88	415.67
2000	$x_{p_1}^*$	12.69	9.80	7.27	5.68	4.65
	OFV*	2,631.32	1,224.45	755.64	554.47	439.05
3000	$x_{p_1}^*$	15.33	10.58	7.60	5.86	4.76
	OFV*	3,107.51	1,307.25	783.73	568.57	447.39
4000	$x_{p_1}^*$	17.03	11.01	7.77	5.96	4.82
	OFV*	3,415.88	1,352.89	798.54	575.88	451.68

It is interesting to note from Table 2.1 that, under a specific unit penalty cost, the path flow will be reduced when the loss of testing and processing, $(1-\alpha_c)$, increases within some specific range. For instance, when λ_1^- is 200 and the loss, $1-\alpha_c$, increases from 0.4 to 0.8, the optimal path flow decreases from 2.83 to 0.88. However, the value of the objective function at the optimal solution always keeps on increasing. Table 2.1 also illustrates, under the same loss associated with testing and processing, the optimal path flow rises with an increase in the unit shortage penalty cost, λ_1^-.

The optimal path flow in Example 2.2, assuming a nonnegative value, as a function of the link multiplier and the shortage penalty *simultaneously* can be expressed as follows:

$$x_{p_1}^* = \frac{5\alpha_c(\lambda_1^- - 14) - 120}{\alpha_c^2(\lambda_1^- + 50) + 65}. \tag{2.56}$$

Of course, for supply chain network topologies of very specialized structure, as the one in Fig. 2.2, one may be able to obtain the optimal solution explicitly using algebra. However, for blood supply chain networks of more general topologies, reflecting additional facilities and activities, as well as additional demand points, an algorithm is needed to determine the optimal solution. We discuss such an algorithm in the next section.

2.3 The Algorithm

In this section, we state the Euler method, which is induced by the general iterative scheme of Dupuis and Nagurney (1993). Its realization for the solution of the blood supply chain network optimization problem governed by variational inequality (2.38) [see also (2.41)] induces subproblems that can be solved explicitly and in closed form. We also use it to compute solutions to other supply chain network problems in Part III of this book. We first recall the following definition.

Definition 2.2 (Norm Projection). Let $\mathcal{K}$ be a closed convex set in R^n. Then for each $Y \in R^n$, there is a unique point $y \in \mathcal{K}$, such that

$$\|Y - y\| \leq \|Y - z\|, \quad \forall z \in \mathcal{K}, \tag{2.57}$$

and y is known as the orthogonal projection of Y on the set $\mathcal{K}$ with respect to the Euclidean norm, that is,

$$y = \Pi_{\mathcal{K}} Y = \arg\min_{z \in \mathcal{K}} \|Y - z\|. \tag{2.58}$$

Specifically, at an iteration τ of the Euler method (see also Nagurney and Zhang 1996), one computes

$$X^{\tau+1} = \Pi_{\mathcal{K}}(X^\tau - a_\tau F(X^\tau)), \tag{2.59}$$

where $\Pi_{\mathscr{K}}$ is the projection on the feasible set $\mathscr{K}$, as in (2.58), and F is the function that enters the variational inequality problem (2.41).

As shown in Dupuis and Nagurney (1993) (see also Nagurney and Zhang 1996), for convergence of the general iterative scheme, which induces the Euler method, among other methods, the sequence $\{a_\tau\}$ must satisfy $\sum_{\tau=0}^{\infty} a_\tau = \infty$, $a_\tau > 0$, $a_\tau \to 0$, as $\tau \to \infty$. Specific conditions for convergence of this scheme can be found for a variety of network-based problems, similar to those constructed here, in Nagurney and Zhang (1996) and the references therein.

2.3.1 Explicit Formulae for the Euler Method Applied to the Blood Supply Chain Network Variational Inequality (2.38)

The elegance of this procedure for the computation of solutions to the blood supply chain network optimization problem modeled in Sect. 2.2 can be seen in the following explicit formulae. In particular, (2.59) for the blood supply chain network problem governed by variational inequality problem (2.38) yields the following closed form expression for the blood product path flow on each path $p \in \mathscr{P}_k; k = 1, \ldots, n_R$, at iteration $\tau + 1$:

$$x_p^{\tau+1} = \max\left\{0, x_p^\tau + a_\tau \left(\lambda_k^- \mu_p \left(1 - P_k\left(\sum_{q \in \mathscr{P}_k} x_q^\tau \mu_q\right)\right) - \lambda_k^+ \mu_p P_k\left(\sum_{q \in \mathscr{P}_k} x_q^\tau \mu_q\right)\right.\right.$$

$$\left.\left. - \frac{\partial \sum_{q \in \mathscr{P}} \hat{C}_q(x^\tau)}{\partial x_p} - \frac{\partial \sum_{q \in \mathscr{P}} \hat{Z}_q(x^\tau)}{\partial x_p} - \theta \frac{\partial \sum_{q \in \mathscr{P}} \hat{R}_q(x^\tau)}{\partial x_p}\right)\right\}. \quad (2.60)$$

Interestingly, the induced expressions in (2.60) correspond to the analogues for the Euler method for traffic network equilibrium problems with elastic demands, with convergence guaranteed under reasonable conditions (cf. Nagurney and Zhang 1996). In our setting, the generalized marginal link cost functions, which consist of the sum of the components of the marginal costs and the weighted marginal risk and are given for link a by $\frac{\partial \hat{c}_a(f_a)}{\partial f_a} + \frac{\partial \hat{z}_a(f_a)}{\partial f_a} + \theta \frac{\partial \hat{r}_a(f_a)}{\partial f_a}$, would have to be strictly monotone increasing and *regular*. By regular is meant that the generalized marginal link cost on each link would approach infinity as the link flow approaches infinity. The *inverse demands*, with the counterpart here being the algebraically simplified three terms following the a_τ term in (2.60), would have to be strictly monotone decreasing. These conditions are satisfied by the functions in our case study. Additional use of strict monotonicity is made in Chap. 4 and that of the less restrictive monotonicity, in Chap. 3, where definitions of these properties are given.

2.4 A Case Study

We now apply the Euler method to compute solutions to a larger-scale numerical blood supply chain network problem, which comprises the case study. The supply chain network consisted of two blood collection sites, two blood centers, two component labs, two storage facilities, two distribution centers, and three demand points, as depicted in Fig. 2.3.

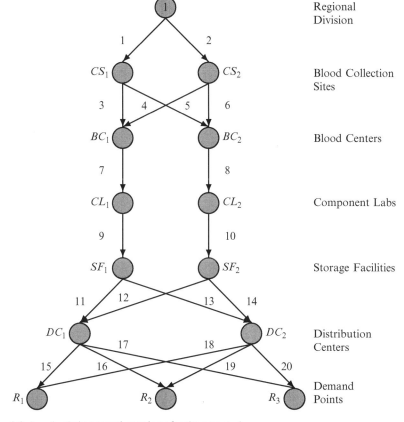

Fig. 2.3 Supply chain network topology for the case study

R_1 was a small surgical center, while R_2 and R_3 were larger hospitals with higher demand for red blood cells. Assuming a weekly schedule for the distribution of product to these demand points, the demands at R_1, R_2, and R_3, over the planning horizon of one week, followed a uniform probability distribution on the intervals [5,10], [40,50], and [25,40], respectively. Hence,

$$P_1\Big(\sum_{p\in\mathcal{P}_1}\mu_p x_p\Big) = \frac{\sum_{p\in\mathcal{P}_1}\mu_p x_p - 5}{5}, \quad P_2\Big(\sum_{p\in\mathcal{P}_2}\mu_p x_p\Big) = \frac{\sum_{p\in\mathcal{P}_2}\mu_p x_p - 40}{10},$$

$$P_3\Big(\sum_{p\in\mathcal{P}_3}\mu_p x_p\Big) = \frac{\sum_{p\in\mathcal{P}_3}\mu_p x_p - 25}{15},$$

where the first O/D pair of nodes was $w_1 = (1,R_1)$, the second was $w_2 = (1,R_2)$, and the third was $w_3 = (1,R_3)$.

The shortage and surplus penalties for each of the three demand points were

$$\lambda_1^- = 2200, \quad \lambda_1^+ = 50, \quad \lambda_2^- = 3000, \quad \lambda_2^+ = 60, \quad \lambda_3^- = 3000, \quad \lambda_3^+ = 50.$$

The total risk functions corresponding to the blood collection links were

$$\hat{r}_1(f_1) = 2f_1^2 \text{ and } \hat{r}_2(f_2) = 1.5f_2^2,$$

the risk-aversion factor, θ, was 0.7, and all other link risk functions were zero.

The multipliers corresponding to the links, the total operational cost functions, and the total discarding cost functions were as reported in Table 2.2. They were determined using the average historical data for the American Red Cross Northeast Division Blood Services (Rios 2010).

The Euler method [cf. (2.60)] for the solution of variational inequality (2.38) was implemented in MATLAB on a PC. We set the sequence $\{a_\tau\} = .1\left(1, \frac{1}{2}, \frac{1}{2}, \dots\right)$, and the convergence tolerance was $\varepsilon = 10^{-6}$, that is, the absolute value of the difference between each path flow in two successive iterations had to be less than or equal to this ε for the implemented algorithm to terminate. The algorithm was initialized by setting the projected demand at each demand point and all other variables equal to zero. Table 2.2 also provides the optimal computed link flow solution.

Table 2.2 Total operational cost and total discarding cost functions and the optimal link flow pattern for the case study

Link a	α_a	$\hat{c}_a(f_a)$	$\hat{z}_a(f_a)$	f_a^*	Link a	α_a	$\hat{c}_a(f_a)$	$\hat{z}_a(f_a)$	f_a^*
1	.97	$6f_1^2+15f_1$	$.8f_1^2$	54.72	11	1.00	$.3f_{11}^2+f_{11}$	$.3f_{11}^2$	29.68
2	.99	$9f_2^2+11f_2$	$.7f_2^2$	43.90	12	1.00	$.5f_{12}^2+2f_{12}$	$.4f_{12}^2$	13.08
3	1.00	$.7f_3^2+f_3$	$.6f_3^2$	30.13	13	1.00	$.4f_{13}^2+2f_{13}$	$.3f_{13}^2$	26.20
4	.99	$1.2f_4^2+f_4$	$.8f_4^2$	22.42	14	1.00	$.6f_{14}^2+f_{14}$	$.4f_{14}^2$	13.31
5	1.00	$f_5^2+3f_5$	$.6f_5^2$	19.57	15	1.00	$1.3f_{15}^2+3f_{15}$	$.7f_{15}^2$	5.78
6	1.00	$.8f_6^2+2f_6$	$.8f_6^2$	23.46	16	1.00	$.8f_{16}^2+2f_{16}$	$.4f_{16}^2$	25.78
7	.92	$2.5f_7^2+2f_7$	$.5f_7^2$	49.39	17	.98	$.5f_{17}^2+3f_{17}$	$.5f_{17}^2$	24.32
8	.96	$3f_8^2+5f_8$	$.8f_8^2$	42.00	18	1.00	$.7f_{18}^2+2f_{18}$	$.7f_{18}^2$	0.29
9	.98	$.8f_9^2+6f_9$	$.4f_9^2$	43.63	19	1.00	$.6f_{19}^2+4f_{19}$	$.4f_{19}^2$	18.28
10	1.00	$.5f_{10}^2+3f_{10}$	$.7f_{10}^2$	39.51	20	.98	$1.1f_{20}^2+5f_{20}$	$.5f_{20}^2$	7.29

2.4 A Case Study

Table 2.3 reports the path flows in the optimal computed path flow pattern. All paths had positive flows for each O/D pair, although several flows for the first O/D pair were close to zero.

Table 2.3 Optimal path flow pattern for the case study

	Path definition	Path flow
O/D pair $w_1 = (1, R_1)$	$p_1 = (1,3,7,9,11,15)$	$x^*_{p_1} = 1.77$
	$p_2 = (1,3,7,9,12,18)$	$x^*_{p_2} = 0.09$
	$p_3 = (1,4,8,10,13,15)$	$x^*_{p_3} = 1.73$
	$p_4 = (1,4,8,10,14,18)$	$x^*_{p_4} = 0.09$
	$p_5 = (2,5,7,9,11,15)$	$x^*_{p_5} = 1.77$
	$p_6 = (2,5,7,9,12,18)$	$x^*_{p_6} = 0.09$
	$p_7 = (2,6,8,10,13,15)$	$x^*_{p_7} = 1.76$
	$p_8 = (2,6,8,10,14,18)$	$x^*_{p_8} = 0.09$
O/D pair $w_2 = (1, R_2)$	$p_1 = (1,3,7,9,11,16)$	$x^*_{p_1} = 12.53$
	$p_2 = (1,3,7,9,12,19)$	$x^*_{p_2} = 5.97$
	$p_3 = (1,4,8,10,13,16)$	$x^*_{p_3} = 7.50$
	$p_4 = (1,4,8,10,14,19)$	$x^*_{p_4} = 5.12$
	$p_5 = (2,5,7,9,11,16)$	$x^*_{p_5} = 4.79$
	$p_6 = (2,5,7,9,12,19)$	$x^*_{p_6} = 5.29$
	$p_7 = (2,6,8,10,13,16)$	$x^*_{p_7} = 6.51$
	$p_8 = (2,6,8,10,14,19)$	$x^*_{p_8} = 6.29$
O/D pair $w_3 = (1, R_3)$	$p_1 = (1,3,7,9,11,17)$	$x^*_{p_1} = 8.74$
	$p_2 = (1,3,7,9,12,20)$	$x^*_{p_2} = 2.29$
	$p_3 = (1,4,8,10,13,17)$	$x^*_{p_3} = 6.69$
	$p_4 = (1,4,8,10,14,20)$	$x^*_{p_4} = 2.21$
	$p_5 = (2,5,7,9,11,17)$	$x^*_{p_5} = 6.48$
	$p_6 = (2,5,7,9,12,20)$	$x^*_{p_6} = 2.18$
	$p_7 = (2,6,8,10,13,17)$	$x^*_{p_7} = 7.66$
	$p_8 = (2,6,8,10,14,20)$	$x^*_{p_8} = 2.38$

The computed optimal amounts of projected demand for the three demand points were $v^*_1 = 6.06$, $v^*_2 = 44.05$, and $v^*_3 = 30.99$. Note that the organization needs to procure an amount $f^*_1 + f^*_2 = 98.62$ with a projected demand of $v^*_1 + v^*_2 + v^*_3 = 81.10$. Because of the product perishability, a higher amount needs to be procured than is projected on the demand side. Hence, the model can assist decision-makers in matching supply with demand under uncertainty and product perishability.

The values of the total operational cost, the total discarding cost, and the weighted total risk at the optimal computed solution were 55,445.55, 11,199.90, and 6,214.72, respectively. The values of the total expected shortage and the total surplus penalties over the three demand points were 16,839.53 and 114.66, respectively. As a result, the value of the objective function (2.17), equivalently, (2.18), at the optimal computed solution, was 89,814.36.

Interestingly, between the two blood collection links, although link 1 has a higher waste/loss rate, and higher total risk and discarding costs, it has a higher optimal flow of blood product as compared to link 2 due to its lower total operational cost function. Furthermore, for the small surgical center, R_1, the value of projected demand, $v_1^* = 6.06$, is closer to the lower bound of its uniform probability distribution due to the relatively smaller shortage penalty cost. In contrast, the values of projected demands for the larger hospitals, R_2 and R_3, are closer to the respective upper bounds of their uniform distributions.

2.5 Summary and Conclusions

In this chapter, we developed a supply chain network optimization model for the management of the procurement, testing and processing, and distribution of a perishable product—that of human blood. For the sake of generality, and the establishment of the foundations that can enable further extensions, adaptations, and applications, some of which are presented in subsequent chapters of this book, we used a variational inequality approach for model formulation and solution. We illustrated the model through transparent numerical examples, which vividly demonstrate the flexibility and generality of our supply chain network optimization model.

The blood supply chain network optimization model in this chapter incorporated demand uncertainty and can provide decision-makers with valuable insights and information, including the blood supply volume that needs to be procured in order to match supply with the demand, in the case of a perishable product. In the next chapter, the generalized network supply chain model for medical nuclear products will have known and fixed demands, since the associated medical diagnostic procedures are scheduled a priori. In Part III of this book, the demands associated with the products will, in turn, be price-sensitive. In Chap. 5, where the focus is pharmaceutical supply chains, we also present corollaries that include fixed and known demands.

2.6 Sources and Notes

The model in this chapter is an extension of the supply chain network model in the paper by Nagurney et al. (2012) and includes risk functions on all the links, and not just on the procurement links. This chapter contains additional theoretical constructs and solution output information to those given in Nagurney et al. (2012). A related paper by Nagurney and Masoumi (2012) further applies some of the theoretical results and concepts discussed in this chapter to the problem of the optimal design of blood supply chain networks.

In this chapter, we have set the foundation for subsequent chapters in this book through the definition of such useful constructs as link/arc multipliers, path multipliers, and arc–path multipliers. In Chap. 3, we demonstrate how radioactive decay over time can be captured in such constructs to model the decay of radioisotopes used in medical diagnostics in the context of another supply chain network optimization model. In Chap. 4, we formulate the decay of fresh produce with the use of such multipliers in a competitive supply chain network.

We briefly highlight other methodological approaches that have been applied to blood banking and related systems. Several authors have applied integer optimization models such as facility location, set covering, allocation, and routing to address the optimization/design of supply chains of blood or other perishable critical products (see Jacobs et al. 1996; Pierskalla 2004; Sahin et al. 2007; Cetin and Sarul 2009; Ghandforoush and Sen 2010). In addition to inventory management methods, noted in Chap. 1, Markov models (Boppana and Chalasani 2007) as well as simulation techniques (cf. Mustafee et al. 2009) have been used to handle blood banking systems. Yegul (2007) also utilized simulation for a blood supply chain with a focus on Turkey. He noted that there were few studies which consider multiple echelons (as the model in this chapter does).

References

Boppana RV, Chalasani S (2007) Analytical models to determine desirable blood acquisition rates. Proceedings of the IEEE international conference on system of systems engineering. San Antonio, Texas, p 6

Cetin E, Sarul LS (2009) A blood bank location model: a multiobjective approach. Eur J Pure Appl Math 2(1):112–124

Chankong V, Haimes YY (1983) Multiobjective decision making: theory and methodology. North-Holland, New York

Chen I (2010) In a world of throwaways, making a dent in medical waste. The New York Times, July 5. Available online at: http://www.nytimes.com/2010/07/06/health/06waste.html

DiBlasio M (2012) Red Cross says blood supply at lowest level in 15 years. USA Today, July 28. Available online at: http://www.usatoday.com/news/health/story\\/2012-07-28/nationwide-blood-shortage/56545264/1

Dong J, Zhang D, Nagurney A (2004) A supply chain network equilibrium model with random demands. Eur J Oper Res 156(1):194–212

Dupuis P, Nagurney A (1993) Dynamical systems and variational inequalities. Ann Oper Res 44:9–42

Fishburn PC (1970) Utility theory for decision making. Wiley, New York

Ghandforoush P, Sen TK (2010) A DSS to manage platelet production supply chain for regional blood centers. Decis Support Syst 50(1):32–42

Hwang H, Hahn KH (2000) An optimal procurement policy for items with an inventory level-dependent demand rate and fixed lifetime. Eur J Oper Res 127(3):537–545

Jacobs DA, Silan MN, Clemson BA (1996) An analysis of alternative locations and service areas of American Red Cross blood facilities. Interfaces 26(3):40–50

Keeney RL, Raiffa H (1993) Decisions with multiple objectives: preferences and value tradeoffs. Cambridge University Press, Cambridge

Mustafee N, Taylor SJE, Katsaliaki K, Brailsford SC (2009) Facilitating the analysis of a UK national blood service supply chain using distributed simulation. Simulation 85(2):113–128

Nagurney A (1999) Network economics: a variational inequality approach, 2nd and revised edn. Kluwer Academic, Dordrecht

Nagurney A, Aronson J (1989) A general dynamic spatial price network equilibrium model with gains and losses. Networks 19(7):751–769

Nagurney A, Cruz J, Dong J, Zhang D (2005) Supply chain networks, electronic commerce, and supply side and demand side risk. Eur J Oper Res 164(1):120–142

Nagurney A, Dong J (2002) Supernetworks: decision-making for the information age. Edward Elgar Publishing, Cheltenham

Nagurney A, Masoumi AH (2012) Supply chain network design of a sustainable blood banking system. In: Boone T, Jayaraman V, Ganeshan R (eds) Sustainable supply chains: models, methods and public policy implications. Springer, London, pp 49–72

Nagurney A, Masoumi AH, Yu M (2012) Supply chain network operations management of a blood banking system with cost and risk minimization. Comput Manag Sci 9(2):205–231

Nagurney A, Qiang Q (2009) Fragile networks: identifying vulnerabilities and synergies in an uncertain world. Wiley, Hoboken, NJ

Nagurney A, Yu M, Qiang Q (2011) Supply chain network design for critical needs with outsourcing. Paper Reg Sci 90(1):123–143

Nagurney A, Zhang D (1996) Projected dynamical systems and variational inequalities with applications. Kluwer Academic, Boston, MA

Nahmias S (1982) Perishable inventory theory: a review. Oper Res 30(4):680–708

Pierskalla WP (2004) Supply chain management of blood banks. In: Brandeau ML, Sanfort F, Pierskalla WP (eds) Operations research and health care: a handbook of methods and applications. Kluwer Academic, Boston, MA, pp 103–145

Redcrossblood.org (2010) Donation FAQs. Available online at: http://redcrossblood.org/donating-blood/donation-faqs

Rios J (2010) Interview with the Medical Director for the American Red Cross Northeast Division Blood Services, Dedham, MA, on July 19, 2010

Sahin G, Sural H, Meral S (2007) Locational analysis for regionalization of Turkish Red Crescent blood services. Comput Oper Res 34(3):692–704

Sullivan MT, Cotten R, Read EJ, Wallace EL (2007) Blood collection and transfusion in the United States in 2001. Transfusion 47(3):385–394

Whitaker BI, Green J, King MR, Leibeg LL, Mathew SM, Schlumpf KS, Schreiber GB (2007) The 2007 national blood collection and utilization survey report. The United States Department of Health and Human Services

Yegul M (2007) Simulation analysis of the blood supply chain and a case study. Master's Thesis, Middle East Technical University, Turkey

Zhou Y-W, Yang S-L (2003) An optimal replenishment policy for items with inventory-level-dependent demand and fixed lifetime under the LIFO policy. J Oper Res Soc 54(6):585–593

Chapter 3
Medical Nuclear Supply Chains

Abstract In this chapter, we present a generalized network model for the optimization of the complex operations of medical nuclear supply chains in the case of the radioisotope molybdenum, with a focus on minimizing the total operational cost, the total waste cost, and the risk associated with this highly time-sensitive and perishable, yet, critical, product used in healthcare diagnostics. Our model allows for the evaluation of transitioning the production and processing of the radioisotope from highly enriched uranium targets to low enriched uranium targets. A case study for North America demonstrates how our framework can be applied in practice.

3.1 Motivation and Overview

In this chapter, we explore medical nuclear supply chain networks, which, as the blood supply chain networks described in Chap. 2, are critical supply chains in healthcare. They provide the conduits for products used in nuclear medical imaging, which is routinely utilized by physicians for diagnostic analysis. Each day, 41,000 nuclear medical procedures are performed in the United States using technetium-99m, a radioisotope obtained from the decay of molybdenum-99. We concentrate on molybdenum-99 due to its importance in medical diagnostics, its time-sensitive nature, and the fact that there are only a handful of production and processing facilities for this radioisotope globally. Such supply chains have unique features and characteristics due to the products' time sensitivity along with their hazardous nature. In this chapter, we construct a model for the optimization of supply chain networks of medical nuclear products, which captures some of the salient issues surrounding such supply chains, from their complexity, to the economic aspects, to the underlying physics of radioactive decay, and the inclusion of waste management.

Molybdenum-99 is produced by primarily irradiating highly enriched uranium (HEU) targets in research reactors. For over two decades, no irradiation and subsequent molybdenum processing has occurred in the United States. All of the molybdenum necessary for US-based nuclear medical diagnostic procedures, which include diagnostics for two of the greatest killers, cancer and cardiac problems, comes from foreign sources. Since molybdenum-99 has a half-life of only 66.7 h, continuous production is needed to provide the supply for the medical procedures. Thus, the USA is critically vulnerable to molybdenum supply chain disruptions that could significantly affect our healthcare security.

Currently, approximately 60% of the supply of molybdenum-99 (^{99}Mo) needed for medical procedures in the United States comes from a Canadian reactor. The remainder comes from Western Europe. There are only nine reactors currently in the world being used for target irradiation and six major processing plants. The shutdown of any of the reactors or processing plants, whether due to routine maintenance, upgrades, or, as occurred during 2009 and 2010, for emergency repairs, could significantly disrupt the molybdenum supply and impact medical facilities' abilities to perform the necessary imaging (see Ponsard 2010).

Additional challenges to the ^{99}Mo supply chain lie in the reality that limitations in processing capabilities restrict the ability to produce the medical radioisotopes from regional reactors since long-distance transportation of the product during certain stages in the supply chain raises safety and security risks and also results in greater decay of the product. There are fewer than a dozen generator manufacturers, which package the radioisotopes into containers known as generators, which are directly used by hospitals and imaging facilities (OECD Nuclear Energy Agency 2010b). In addition, since the majority of the reactors are between 40 and 50 years old, several of the reactors currently used, including the Canadian one, are scheduled to be retired by the end of this decade (Seeverens 2010; OECD Nuclear Energy Agency 2010a).

A proper model of this critical medical nuclear supply chain, which allows for appropriate economic cost quantification, heavily emphasized by policy-makers (see OECD Nuclear Energy Agency 2011), must include the physics-based principles of the underlying radioactivity. It must also incorporate multicriteria decision-making and optimization to capture the operational and waste management costs as well as risk management, subject to constraints of demand satisfaction at the hospitals and medical facilities. Moreover, the model must be sufficiently flexible and robust in order to provide rigorous solutions as the technological landscape changes.

In Sect. 3.2, we develop the supply chain network optimization model for molybdenum, ^{99}Mo, and describe the various tiers of the supply chain network, consisting of the nuclear reactors, the processors, the generator manufacturing facilities, and, finally, the hospitals and medical facilities, where the medical radioisotopes are injected in the patients. As we did for the blood supply chain model in Chap. 2, we model the medical nuclear supply chain network optimization problem as a multicriteria system-optimization problem on a generalized network. We identify the specific losses on the links/arcs through the use of the time decay of the radioisotope, and we capture distinctions between LEU (low enriched uranium) versus HEU target irradiation and processing. We consider total cost minimization associated with the operational costs, along with the waste management costs, since we are dealing with nuclear products, and the associated risk of the various supply chain network activities which is especially relevant here since these are hazardous products and by-products. Medical nuclear waste management issues have not received much attention in recent reports (cf. OECD Nuclear Energy Agency 2010a,b). Following Alp (1995), we define risk as a measure of the probability and the severity of harm due to the release and disposal of the associated materials. In the processes of manufacturing and of distribution/storage, the risk functions are concentrated on the quantity of the medical nuclear product, which is captured by the link flow. In transportation/shipment, the risk functions depend on the travel time and the accident probabilities, which are also realized by the link flows and are coupled with the transportation modes. In the processes of manufacturing and distribution/storage, risks are associated with the production and the disposal of hazmat. Although there are no risk assessment functions that are well established and widely applied in the literature of medical nuclear transportation, various risk functions have been presented to estimate the risk in hazmat transportation.

In Sect. 3.3, we propose a computational approach, which resolves the medical nuclear supply chain network variational inequality reformulation of the optimization problem into subproblems that can be solved explicitly and exactly at each iteration. In Sect. 3.4, we compute solutions to a realistic medical nuclear supply chain case study. In Sect. 3.5, we summarize our findings and present our conclusions. Section 3.6 contains the "Sources and Notes."

3.2 The Medical Nuclear Supply Chain Network Model

In this section, we briefly include some background on medical nuclear technology. To produce a medical image, a radioactive isotope is bound to a pharmaceutical that is injected into the patient and travels to the site or organ of interest. The gamma rays emitted by the radioactive decay of the isotope are then used to construct an image of that site or organ (Berger et al. 2004). Technetium, ^{99m}Tc, which is a decay product of molybdenum-99, ^{99}Mo, is the most commonly used medical radioisotope, used in more than 80% of the radioisotope injections, with more than 30 million procedures worldwide each year. Over 100,000 medical facilities in the world use radioisotopes (World Nuclear Association 2012). In 2008, 18.5 million doses of ^{99m}Tc were injected in the USA with 2/3 of them used for cardiac exams (Lantheus Medical Imaging, Inc. 2009). By using medical radioisotope techniques, health professionals can enable the earlier and more accurate detection of cardiac problems and cancer, the two most common causes of death (see Kochanek et al. 2011). According to Kahn (2008), the global market for medical isotopes is 3.7 billion US$ per year.

The technology and policy landscape is now changing for medical nuclear supply chains. Although currently most of the ^{99}Mo production uses HEU targets, all producing countries, where economically and technologically feasible, have agreed, in principle, to convert to low enriched uranium (LEU) according to an OECD Nuclear Energy Agency (2011) report. The use of LEU targets for ^{99}Mo production has several advantages over HEU, primarily due to proliferation resistance and, hence, enhanced global security, the easier availability of the target material, and also easier compliance for its transportation and processing. Its negatives include, however, a lower production yield than HEU and a greater number of targets that are needed to be irradiated with associated increased volumes of waste. Indeed, according to Kramer (2011), the South African Nuclear Energy Corp (Necsa) believes that the LEU production process will approximately double the amount of waste generated in extracting the radioisotope, whereas other producers are likely to see a factor-of-four increase in their wastes. Hence, both production and processing pressures are raised as well as waste management issues.

Since ^{99}Mo decays with a 66.7 h half-life, approximately 99.9% of the atoms decay in 27.5 days, making its production, transportation, and processing all extremely time-sensitive. In fact, the production of ^{99}Mo is quantified in *six-day curies end of processing* denoting the activity of the sample 6 days after it was irradiated to highlight this (see OECD Nuclear Energy Agency 2010a). In addition to the time-sensitivity, the irradiated targets are highly radioactive, restricting HEU target transportation options between the reactor and the processing facilities to only trucks that can transport the heavily shielded transportation containers. While the extracted bulk ^{99}Mo continues to be constrained by its decay, its shielding requirements are reduced, allowing for transportation by multiple modes, including by air (cf. de Lange 2010). Irradiated LEU targets, however, can be transported by plane, opening up an alternative transportation option to that of trucks, with implications for the medical nuclear supply chain.

In this section, we develop the medical nuclear supply chain network optimization model, with a focus on ^{99}Mo, referred to, henceforth, as *Mo*. We note that the framework is also relevant, with minor modifications, to other radioisotopes, including iodine-131. Figure 3.1 depicts the network topology of the medical nuclear supply chain. In the network, the top level (origin) node 1 represents the organization and the bottom level nodes represent the destination nodes. Every other node in the network denotes a component/facility in the medical nuclear system. A path connecting the origin node to a destination node, corresponding to a demand point, consists of a sequence of directed links which correspond to the supply chain network activities that guarantee that the nuclear product is produced, processed, and, ultimately, distributed to the hospitals and medical imaging facilities, where it is administered to the patients. It is assumed that there exists at least one path joining node 1 with each destination node: $H_1^2, \ldots, H_{n_H}^2$. An origin/destination (O/D) pair of nodes w_k for the supply chain in Fig. 3.1 is defined by the pair of nodes $(1, H_k^2)$.

In the network in Fig. 3.1, there are n_R reactor sites, which produce the radioisotope. These are usually government research reactors and are represented by the second tier of nodes of the network: $R_1, R_2, \ldots, R_{n_R}$.

The first set of links, connecting the origin node to the second tier, corresponds to the process of radioisotope production. Different reactors irradiate different targets, that is, either HEU or LEU, and hence, the reactors are irradiation target specific.

The next set of nodes, located in the third tier of the supply chain network in Fig. 3.1, consists of the radioisotope processing centers.

3.2 The Medical Nuclear Supply Chain Network Model

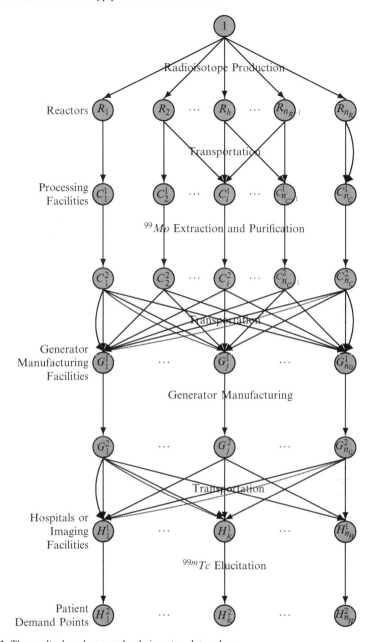

Fig. 3.1 The medical nuclear supply chain network topology

There exist n_C of these facilities, denoted, respectively, by $C_1^1, C_2^1, \ldots, C_{n_C}^1$, to which the irradiated targets containing the Mo are shipped after their irradiation at the reactor sites. Thus, the set of links connecting tiers two and three of the network topology represents the transportation of the radioisotope. Transportation of HEU targets, which are hazardous materials, is done solely by truck using specialized containers. This single mode of transportation is represented by single links joining the pairs of nodes. Because of this, the HEU processing facilities must be located fairly close to the reactors. However, for LEU targets, either truck or air transport (or both) may be feasible, reflected by multiple links between a reactor and a processor in Fig. 3.1.

At the processing centers, the Mo is extracted and purified. Note that, as depicted by the supply chain network topology, not every processing facility can process both LEU and HEU irradiated targets, but some may be able to, and we consider such a case study in Sect. 3.4. Consequently, in the network in Fig. 3.1, the first processor only processes the type of target produced by the first reactor, and so on. This processing is represented by the links emanating from the nodes $C_1^1, C_2^1, \ldots, C_{n_C}^1$ and ending in the nodes $C_1^2, C_2^2, \ldots, C_{n_C}^2$, with the latter set of nodes being the fourth tier nodes. The last processor in Fig. 3.1 is an LEU processor, and hence, there are multiple transportation options, depicted accordingly by multiple links. Figure 3.1 can be modified as the technology landscape changes during the transition from HEU to LEU target irradiation.

The fifth tier of the network is associated with the generator manufacturing facilities, and these nodes are joined with the fourth tier nodes by links which represent the multiple modes of transportation that are available for transporting the purified Mo to the generator manufacturing facilities. The number of generator manufacturing facilities is given by n_G. These facilities are denoted by $G_1^1, \ldots, G_{n_G}^1$, respectively, and need not be located near the processing facilities.

At the generator manufacturing facilities, the radioisotope is processed and packaged in a form to be used at the medical facilities. The links that emanate from the generator manufacturing facility nodes terminate in the sixth tier set of nodes, respectively, denoted by $G_1^2, \ldots, G_{n_G}^2$ in Fig. 3.1, which represent the completion of this stage of processing. From the latter generator nodes, there emanate transportation links, and these links, as the preceding transportation links, correspond to multiple modes of transportation, as appropriate, including not only trucking but also common carrier air transportation. These links terminate in the seventh tier of nodes, $H_1^1, \ldots, H_{n_H}^1$, which represent the hospitals and the medical facilities that dispense the radioisotope to the patients. There is still one more stage of processing, eluding the generators to produce the final injectable radioisotope. This is represented by the final set of links in Fig. 3.1 terminating in nodes $H_1^2, \ldots, H_{n_H}^2$, which represent the final patient demand points (the patients).

The medical nuclear supply chain network topology, as depicted in Fig. 3.1, is denoted by $\mathscr{G} = [N, L]$, where N and L denote the sets of nodes and links, respectively.

We use a multicriteria system-optimization formalism, since the organization wishes to determine at what level the reactors, the processing centers, and the generator manufacturers should operate. Furthermore, the organization seeks to minimize

3.2 The Medical Nuclear Supply Chain Network Model

the total risk, because it is dealing with a hazardous product, as well as the total operational costs associated with the production, processing, and transportation activities, and the total cost of discarding the nuclear waste product associated with each of the links.

We assume that the demands must be satisfied since many medical procedures using radioisotopes must be scheduled in advance.

The links are denoted by a, b, etc. We associate a unit operational cost function with each link in the supply chain network that reflects the cost of operating that particular medical nuclear supply chain activity. The unit operational cost on link a is denoted by c_a and is a function of flow on that link, f_a. The *total* operational cost on link a is denoted by $\hat{c}_a$ and is constructed as

$$\hat{c}_a(f_a) = f_a \times c_a(f_a), \qquad \forall a \in L. \tag{3.1}$$

The link total cost functions are assumed to be convex and continuously differentiable.

The origin/destination (O/D) nodes (cf. Fig. 3.1) are the pairs of nodes: $(1, H_k^2)$; $k = 1, \ldots, n_H$. Let $\mathscr{P}_k$ denote the set of paths, which represent the alternative associated possible supply chain network processes, joining $(1, H_k^2)$. $\mathscr{P}$ denotes the set of all paths joining node 1 to the destination nodes, and $n_{\mathscr{P}}$ denotes the number of paths.

Also, d_k denotes the demand for the radioisotope at the demand point H_k^2; $k = 1, \ldots, n_H$.

As in Chap. 2, we, again, associate with every link a in the supply chain network an arc multiplier α_a, which corresponds to the percentage of decay and additional loss over that link. This multiplier lies in the range $(0,1]$, for the network activities, where $\alpha_a = 1$ means that 100% of the initial flow on link a reaches the successor node of that link, reflecting that there is no decay/waste/loss on link a. As will be described formally later, the multiplier α_a can be modeled as the product of two terms, a radioactive decay multiplier α_{d_a} and a processing loss multiplier α_{l_a}.

3.2.1 The Underlying Physics

For completeness and easy reference, we now outline the underlying physics of radioactive decay in this application, along with how we model it through arc and path multipliers. The activity of a radioisotope (in disintegrations per unit time) is proportional to the quantity of that isotope, that is,

$$\frac{dN}{dt} \propto N, \tag{3.2}$$

where $N = N(t)$ = the quantity of a radioisotope. The quantity of a radioisotope in a time interval t is then given by

$$N(t) = N_0 e^{-\lambda t}, \tag{3.3}$$

where N_0 is the quantity present at the beginning of the interval and λ is the decay constant of the radioisotope (see Berger et al. 2004).

We can represent the radioactive decay multiplier α_{da} for link a as

$$\alpha_{da} = e^{-\lambda t_a}, \tag{3.4}$$

where t_a is the time spent on link a. The decay constant, λ, in turn, can be conveniently represented by an experimentally measured value, called the half-life, $t_{1/2}$, where

$$t_{1/2} = \frac{\ln 2}{\lambda}. \tag{3.5}$$

The values of the half-lives of radioisotopes are tabulated in the handbook of the American Institute of Physics (1972). Hence, we can write α_{da} as

$$\alpha_{da} = e^{-\lambda t_a} = e^{-\ln 2 \frac{t_a}{t_{1/2}}} = 2^{-\frac{t_a}{t_{1/2}}}. \tag{3.6}$$

The value of $t_{1/2}$ for Mo, as noted in Sect. 3.1, is 66.7 h.

The processing loss multiplier α_{la} for link a is a factor in the range (0,1] that quantifies the losses that occur during processing. Different processing links may have different values for this parameter. For transportation links, however, there are no losses beyond those due to radioactive decay. Therefore, for transportation links, we have that $\alpha_{la} = 1$. We define for every link a the multiplier $\alpha_a = \alpha_{la} \times \alpha_{da}$. For the top-most manufacturing links, we also have that $\alpha_a = 1$.

As mentioned earlier, f_a denotes the (initial) flow on link a. Let f'_a denote the final flow on that link, that is, the flow that reaches the successor node of the link. Therefore,

$$f'_a = \alpha_a f_a, \quad \forall a \in L. \tag{3.7}$$

Since α_a is constant, and known a priori, a total discarding cost function, $\hat{z}_a$, can be defined accordingly. As also done in Chap. 2, it is a function of the flow, f_a, is assumed to be convex and continuously differentiable, and is given by

$$\hat{z}_a = \hat{z}_a(f_a), \quad \forall a \in L. \tag{3.8}$$

Note that, in processing/producing an amount of radioisotope f_a, one knows from the physics the amount of hazardous waste, and hence, a discarding function of the form (3.8) is appropriate. As noted earlier in this section, LEU target processing can generate several times the amount of waste that HEU target processing does (Kramer 2011).

Let x_p represent the (initial) flow of Mo on path p joining the origin node with a destination node. The path flows must be nonnegative, that is,

$$x_p \geq 0, \quad \forall p \in \mathscr{P}, \tag{3.9}$$

since a physical nuclear product will be produced, processed, transported, etc.

3.2 The Medical Nuclear Supply Chain Network Model

Similar to the discussion in Sect. 2.2.2, let μ_p denote the path multiplier corresponding to the loss on path p, which is defined as the product of all link multipliers on links comprising that path, that is,

$$\mu_p \equiv \prod_{a \in p} \alpha_a, \quad \forall p \in \mathscr{P}. \tag{3.10}$$

The demand at demand point H_k, d_k, is known and given and must be equal to the sum of all the final flows on paths joining $(1, H_k^2)$:

$$d_k = \sum_{p \in \mathscr{P}_k} \mu_p x_p, \quad k = 1, \ldots, n_H. \tag{3.11}$$

Indeed, although the amount of radioisotope that originates on a path p is x_p, the amount (due to radioactive decay, etc.) that actually arrives at the destination (terminal node) of this path is $x_p \mu_p$. The ultimate demand is the amount of ^{99m}Tc that is needed for patient procedures at patient demand points $H_1^2, \ldots, H_{n_H}^2$ which is usually expressed in number of curies used per week as further elaborated upon in the numerical case study in Sect. 3.4.

The arc–path multiplier α_{ap} is the product of the multipliers of the links on path p that precede link a in that path, as defined in (2.13), and hence, the relationship between the link flow, f_a, and the path flows is

$$f_a = \sum_{p \in \mathscr{P}} x_p \alpha_{ap}, \quad \forall a \in L. \tag{3.12}$$

The total cost minimization objective faced by the organization includes the total cost of operating the various links and the total discarding cost of waste/loss over the links. This optimization problem can be expressed as

$$\text{Minimize} \quad \sum_{a \in L} \hat{c}_a(f_a) + \sum_{a \in L} \hat{z}_a(f_a) \tag{3.13}$$

subject to constraints (3.9), (3.11), and (3.12), and

$$f_a \leq \bar{u}_a, \quad \forall a \in L. \tag{3.14}$$

Constraint (3.14) guarantees that the flow on a link cannot exceed the capacity on that link. Note that no such constraint was used for the blood supply chain network model in Chap. 2.

As mentioned earlier, the minimization of total costs is not the only objective. A major challenge for a medical nuclear organization is to capture the risk associated with different activities in the nuclear supply chain network. This was also an issue for the blood banking system organization in Chap. 2 for its supply chain.

Let $\hat{r}_a$ denote the total risk on link $a \in L$. We assume that the total risk on a link is denoted by a risk function that is a function of the flow on the link, that is,

$$\hat{r}_a = \hat{r}_a(f_a), \quad \forall a \in L. \tag{3.15}$$

We assume that the total risk functions are convex and continuously differentiable.

For example, for a transportation link a, the total risk function would measure the impact of the travel time, the population density that the transportation route goes through, the unit probability of an accident using the particular mode represented by the link, the area of the impact zone, the length of the link, etc., and, ideally, also include the impact of human factors. In the case of a non-transportation processing link, the function would capture analogous aspects but with a focus on the specific processing activity.

The organization attempts to minimize the total risk over all links, similar to the objective of the blood banking organization in Chap. 2. Thus, the risk minimization objective function for the organization can be expressed as

$$\text{Minimize} \quad \sum_{a \in L} \hat{r}_a(f_a). \tag{3.16}$$

Specific functional forms of the total cost functions (3.1) and (3.8) and the total risk functions (3.15) for a numerical case study are described in detail in Sect. 3.4.

3.2.2 The Multicriteria Decision-Making Problem in Link Flows and in Path Flows

The supply chain network optimization problem for a medical nuclear product can be expressed as a multicriteria decision-making problem. The organization seeks to determine the optimal levels of radioisotope processed on each supply chain network link subject to the minimization of the total cost (operational and discarding) as well as the minimization of the total risk. A weight of ω is assigned by the decision-maker to the total risk objective (3.16), and it can be interpreted as the factor of risk aversion. The ω can also be interpreted as a risk to cost conversion factor.

Thus, the multicriteria optimization problem is

$$\text{Minimize} \quad \sum_{a \in L} \hat{c}_a(f_a) + \sum_{a \in L} \hat{z}_a(f_a) + \omega \sum_{a \in L} \hat{r}_a(f_a) \tag{3.17}$$

subject to constraints (3.9), (3.11), (3.12), and (3.14).

The above optimization problem is in terms of link flows. It can also be expressed in terms of path flows:

$$\text{Minimize} \quad \sum_{p \in \mathcal{P}} \left(\hat{C}_p(x) + \hat{Z}_p(x) \right) + \omega \sum_{p \in \mathcal{P}} \hat{R}_p(x) \tag{3.18}$$

subject to constraints (3.9), (3.11), and (3.14) (modified with the use of (3.12) for the link flow(s)). Each path total operational cost function, $\hat{C}_p(x)$; each path total discarding cost function, $\hat{Z}_p(x)$; and each path total risk function, $\hat{R}_p(x)$, are, respectively, derived from $C_p(x)$, $Z_p(x)$, and $R_p(x)$ and are similar to the respective ones in Chap. 2, as follows:

3.2 The Medical Nuclear Supply Chain Network Model

$$\hat{C}_p(x) = x_p \times C_p(x), \hat{Z}_p(x) = x_p \times Z_p(x), \hat{R}_p(x) = x_p \times R_p(x), \quad \forall p \in \mathcal{P}, \quad (3.19)$$

with the unit cost functions on path p, that is, $C_p(x)$, $Z_p(x)$, and $R_p(x)$, in turn, defined as

$$C_p(x) \equiv \sum_{a \in L} c_a(f_a)\alpha_{ap}, Z_p(x) \equiv \sum_{a \in L} z_a(f_a)\alpha_{ap}, R_p(x) \equiv \sum_{a \in L} r_a(f_a)\alpha_{ap}, \quad \forall p \in \mathcal{P}.$$
(3.20)

We associate the Lagrange multiplier γ_a with constraint (3.14) for each link a, and we denote the optimal Lagrange multiplier by $\gamma_a^*, \forall a \in L$. The Lagrange multipliers may be interpreted as shadow prices. We group these Lagrange multipliers into the vector γ.

Let K denote the feasible set such that

$$K \equiv \{(x,\gamma) | x \in R_+^{n_p}, (3.11) \text{ and } (3.14) \text{ with the substitution } (3.12) \text{ hold}, \gamma \in R_+^{n_L}\}.$$
(3.21)

We assume that the feasible set is nonempty.

We now derive the variational inequality formulations of the problem in terms of path flows and link flows.

Theorem 3.1 (Variational Inequality Formulations). *The optimization problem (3.18), subject to its constraints, is equivalent to the variational inequality problem: determine the vector of optimal path flows and the vector of optimal Lagrange multipliers $(x^*, \gamma^*) \in K$, such that*

$$\sum_{k=1}^{n_H} \sum_{p \in \mathcal{P}_k} \left[\frac{\partial (\sum_{q \in \mathcal{P}} \hat{C}_q(x^*))}{\partial x_p} + \frac{\partial (\sum_{q \in \mathcal{P}} \hat{Z}_q(x^*))}{\partial x_p} + \sum_{a \in L} \gamma_a^* \alpha_{ap} + \omega \frac{\partial (\sum_{q \in \mathcal{P}} \hat{R}_q(x^*))}{\partial x_p} \right]$$

$$\times [x_p - x_p^*] + \sum_{a \in L} \left[\bar{u}_a - \sum_{p \in \mathcal{P}} x_p^* \alpha_{ap} \right] \times [\gamma_a - \gamma_a^*] \geq 0, \quad \forall (x, \gamma) \in K. \quad (3.22)$$

The variational inequality (3.22), in turn, can be rewritten in terms of link flows as follows: determine the vector of optimal link flows and the vector of optimal Lagrange multipliers $(f^*, \gamma^*) \in K^1$, such that

$$\sum_{a \in L} \left[\frac{\partial \hat{c}_a(f_a^*)}{\partial f_a} + \frac{\partial \hat{z}_a(f_a^*)}{\partial f_a} + \gamma_a^* + \omega \frac{\partial \hat{r}_a(f_a^*)}{\partial f_a} \right] \times [f_a - f_a^*]$$

$$+ \sum_{a \in L} [\bar{u}_a - f_a^*] \times [\gamma_a - \gamma_a^*] \geq 0, \quad \forall (f, \gamma) \in K^1, \quad (3.23)$$

where K^1 denotes the feasible set:

$$K^1 \equiv \{(f, \gamma) | \exists x \geq 0, (3.11), (3.12), \text{ and } (3.14) \text{ hold, and } \gamma \geq 0\}. \quad (3.24)$$

Proof. First, we prove the result for path flows [cf. (3.22)].

The convexity of $\hat{C}_p$, $\hat{Z}_p$, and $\omega \hat{R}_p$ for all paths p holds since $\hat{c}_a$, $\hat{z}_a$, and $\hat{r}_a$ were assumed to be convex for all links a and ω are nonnegative. Hence, the sum of the path marginal total costs and the path marginal total risks on all paths, as in (3.18), is convex. Since the objective function (3.18) is convex and the feasible set K is closed and convex, the variational inequality (3.22) follows from the standard theory of variational inequalities.

As for the proof of the variational inequality (3.23), now that (3.22) is established, we use Lemma 2.1 to obtain the equivalence between the partial derivatives of the total costs on paths and those of the total risks and the partial derivatives of the total costs and the total risks, respectively, on links. Also, using (3.12) and (3.14), we can rewrite the formulation in terms of link flows rather than path flows. Thus, the second part of Theorem 3.1, that is, the variational inequality in link flows (3.23), also holds. □

Variational inequality (3.22) can be put into standard form VI $(F, \mathcal{K})$ [cf. (2.41)] if we define the feasible set $\mathcal{K} \equiv K$, the vector $X \equiv (x, \gamma)$, and $F(X) \equiv (F_1(X), F_2(X))$, where

$$F_1(X) \equiv \left[\frac{\partial (\sum_{q \in \mathcal{P}} \hat{C}_q(x))}{\partial x_p} + \frac{\partial (\sum_{q \in \mathcal{P}} \hat{Z}_q(x))}{\partial x_p} + \sum_{a \in L} \gamma_a \alpha_{ap} + \omega \frac{\partial (\sum_{q \in \mathcal{P}} \hat{R}_q(x))}{\partial x_p}; p \in \mathcal{P} \right],$$

$$F_2(X) \equiv \left[\bar{u}_a - \sum_{p \in \mathcal{P}} x_p \alpha_{ap}; \quad a \in L \right]. \tag{3.25}$$

3.3 The Computational Approach

In this section, we describe the computational approach for the solution of the medical nuclear supply chain network problem, governed by variational inequality (3.22). Specifically, we propose to use the modified projection method (Korpelevich 1977), where for (3.22) the variables are path flows, rather than link flows (see, e.g., Nagurney and Qiang 2009). This algorithm, in the context of the medical nuclear supply chain network model, yields subproblems, at each iteration, that can be solved exactly, and in closed form, for the path flows, using a variant of the exact equilibration algorithm, adapted to incorporate arc/path multipliers. It also yields, at each iteration, explicit formulae for the Lagrange multipliers. The modified projection is guaranteed to converge if the function F that enters the variational inequality satisfies monotonicity and Lipschitz continuity and that a solution exists. For completeness, we now provide the definitions of these properties. Monotonicity plays a role in variational inequalities similar to that of convexity in optimization problems.

Definition 3.1 (Monotonicity). $F(X)$ is monotone on $\mathcal{K}$ if

$$\langle (F(X^1) - F(X^2)), X^1 - X^2 \rangle \geq 0, \quad \forall X^1, X^2 \in \mathcal{K}. \tag{3.26}$$

3.3 The Computational Approach

Definition 3.2 (Lipschitz Continuity). $F(X)$ is Lipschitz continuous if there exists an $\mathsf{L} > 0$, such that

$$\|F(X^1) - F(X^2)\| \leq \mathsf{L}\|X^1 - X^2\|, \quad \forall X^1, X^2 \in \mathcal{K}. \tag{3.27}$$

Monotonicity follows for the model under our imposed assumptions, and Lipschitz continuity will also hold provided that the marginal total cost and marginal risk functions have bounded second-order partial derivatives.

We now recall the modified projection method, where τ denotes an iteration.

Step 0: Initialization

Set $X^0 \in \mathcal{K}$. Let $\tau = 1$ and let η be a scalar such that $0 < \eta \leq \frac{1}{\mathsf{L}}$, where L is the Lipschitz continuity constant.

Step 1: Computation

Compute $\tilde{X}^\tau$ by solving the VI subproblem:

$$\langle \tilde{X}^\tau + \eta F(X^{\tau-1}) - X^{\tau-1}, X - \tilde{X}^\tau \rangle \geq 0, \quad \forall X \in \mathcal{K}. \tag{3.28}$$

Step 2: Adaptation

Compute X^τ by solving the VI subproblem:

$$\langle X^\tau + \eta F(\tilde{X}^\tau) - X^{\tau-1}, X - X^\tau \rangle \geq 0, \quad \forall X \in \mathcal{K}. \tag{3.29}$$

Step 3: Convergence Verification

If $\max |X_l^\tau - X_l^{\tau-1}| \leq \varepsilon$, for all l, with $\varepsilon > 0$, a prespecified tolerance, then stop; else, set $\tau =: \tau + 1$, and go to Step 1.

The VI subproblems in (3.28) and (3.29) are quadratic programming problems with special structure that result in straightforward computations. Explicit formulae for (3.28) for the medical nuclear supply chain network problem are now given for the Lagrange multipliers. Analogous formulae for (3.29) can then be easily obtained. Subsequently, we follow up with how the path flow values in (3.28) can be determined [a similar approach can then be used to determine the path flows for (3.29)].

3.3.1 Explicit Formulae for the Lagrange Multipliers at Step 1 [cf. (3.28)]

$$\tilde{\gamma}_a^\tau = \max\{0, \gamma_a^{\tau-1} + \eta(\sum_{p \in \mathcal{P}} x_p^{\tau-1} \alpha_{ap} - \bar{u}_a)\}, \quad \forall a \in L. \tag{3.30}$$

Recall that the feasible set $\mathcal{K}$, in terms of the path flows, requires that the path flows be nonnegative and that the demand constraint (3.11) holds for each demand point. The induced path flow subproblems in (3.28) and (3.29), hence, have a special network structure of the form given in Fig. 3.2.

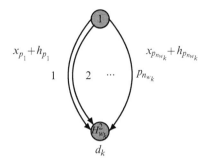

Fig. 3.2 Special network structure of an induced path flow subproblem for each demand point k

Specifically, the path flow subproblems that one must solve in Step 1 [see (3.28)] (we have suppressed the iteration superscripts below) have the following form for each demand point k; $k = 1, \ldots, n_H$:

$$\text{Minimize} \quad \frac{1}{2} \sum_{p \in \mathcal{P}_k} x_p^2 + \sum_{p \in \mathcal{P}_k} h_p x_p \qquad (3.31)$$

subject to

$$d_k = \sum_{p \in \mathcal{P}_k} \mu_p x_p, \qquad (3.32)$$

$$x_p \geq 0, \quad \forall p \in \mathcal{P}_k, \qquad (3.33)$$

where

$$h_p \equiv x_p^{\tau-1} - \eta \left[\frac{\partial (\sum_{q \in \mathcal{P}} \hat{C}_q(x^{\tau-1}))}{\partial x_p} + \frac{\partial (\sum_{q \in \mathcal{P}} \hat{Z}_q(x^{\tau-1}))}{\partial x_p} \right. \\ \left. + \sum_{a \in L} \gamma_a^{\tau-1} \alpha_{ap} + \omega \frac{\partial (\sum_{q \in \mathcal{P}} \hat{R}_q(x^{\tau-1}))}{\partial x_p} \right].$$

The above problem is a quadratic programming problem with special network structure for each demand point k.

We now present an exact equilibration algorithm, adapted to handle the arc and path flow multipliers, which can be applied to compute the solution to problem (3.31) for each demand point, subject to constraints (3.32) and (3.33). An analogous set of subproblems in path flows can be set up and solved accordingly for Step 2 [cf. (3.29)]. For further background on such algorithms, see Dafermos and Sparrow (1969) and Nagurney and Qiang (2009).

3.3.2 An Exact Equilibration Algorithm for a Specially Structured Generalized Network

The statement of the algorithm, which is embedded in the modified projection method to compute the solution to the just described quadratic programming problem for each demand point, is detailed below.

Step 0: Sort

Sort the fixed terms $\frac{h_p}{\mu_p}$; $p \in \mathscr{P}_k$ in nondescending order and relabel the paths/links accordingly. Assume, from this point on, that they are relabeled. Set $r = 1$.

Step 1: Computation

Compute

$$\lambda_k^r = \frac{\sum_{i=1}^r \mu_{p_i} h_{p_i} + d_k}{\sum_{i=1}^r \mu_{p_i}^2}. \tag{3.34}$$

Step 2: Evaluation

If $\frac{h_{p_r}}{\mu_{p_r}} < \lambda_k^r \leq \frac{h_{p_{r+1}}}{\mu_{p_{r+1}}}$, then stop; set $s = r$ and go to Step 3 below; otherwise, let $r = r + 1$ and return to Step 1 above. If $r = n_k$, where n_k denotes the number of paths connecting destination node H_k^2 with origin node 1, then set $s = n_k$ and go to Step 3 below.

Step 3: Path Flow Determination

Set

$$x_{p_i} = \mu_p \lambda_k^s - h_{p_i}, \quad i = 1, \ldots, s.$$
$$x_{p_i} = 0, \quad i = s+1, \ldots, n_k. \tag{3.35}$$

3.4 A Case Study

In this section, we describe a case study. In particular, we consider the molybdenum-99 supply chain in North America, depicted in Fig. 3.3, including the existing Canadian reactor, NRU, the AECL-MDS Nordion processing facility located in Ottawa, and the two US generator manufacturing facilities. This reactor and the processing facility are likely to be decommissioned around 2016. The NRU reactor is located in Chalk River, Ontario, and uses HEU targets. Transportation of the irradiated targets from the reactor to the processing facility takes place by truck. There are two generator manufacturers in the United States (and none in Canada) located in Billerica, Massachusetts, and outside of St. Louis, Missouri.

In addition, in this case study, we have included another potential source of irradiated targets. As discussed in Kramer (2011), Canada's TRIUMF linear accelerator located in Vancouver, British Columbia, is expected to produce molybdenum, irradiated from LEU targets, at TRIUMF in 2012. Hence, we are interested in optimizing the operations of this medical nuclear supply chain scenario. During the transition from NRU HEU targets to TRIUMF LEU, the Nordion processing facility will be able to process both HEU and LEU targets, as depicted in Fig. 3.3. Hence, each of these target processing stages has its own processing link (cf. links 3 and 24 in Fig. 3.3). There are also expected to be available two modes of transportation from TRIUMF to Nordion, truck and air transport, as depicted by two associated alternative transportation links (22, 23) in Fig. 3.3.

Links 4, 6, 9, 11, 13, 14, 16, 17, and 23 correspond to transportation by air, whereas links 2, 5, 10, 12, 15, and 22 correspond to transportation by truck. The capacities on the links were obtained from OECD Nuclear Energy Agency (2010b, 2011) and Kramer (2011). The transportation links are assumed not to be capacity limited (i.e., we assume very large capacities), which we denote by *large* in Table 3.1. In the computations, we set the value to 5,000,000 for all such links. We implemented the modified projection method, along with the generalized exact equilibration algorithm, as described in Sect. 3.3. The ε in the convergence criterion was 10^{-6} [see following (3.29)]. The algorithm was implemented in FORTRAN, and a Linux-based system was used for the computations.

We calculated the values of the arc multipliers α_{da}, for all links $a = 1, \ldots, 24$, using data in the OECD Nuclear Energy Agency (2010a) report and in the National Research Council (2009) report, which included the approximate times associated with the various links in the supply chain network in Fig. 3.3. According to OECD Nuclear Energy Agency (2010a), we may assume that there is no loss a_{la} on each link a for $a = 1, \ldots, 24$, except for processing links 3 and 24. Hence, $\alpha_{la} = 1$ for all the former links; therefore, $\alpha_a = \alpha_{da}$ for all those links, as reported in Table 3.1. In the case of links 3 and 24, $\alpha_{la} = .8$ and $\alpha_{da} = .883$; therefore, $\alpha_3 = \alpha_{24} = .706$. All flows (and capacities) are reported in curies.

Operating cost data were taken from OECD Nuclear Energy Agency (2010b) and converted to per curie processed or generated. As noted by the National Research Council (2009), the US generator prices are proprietary but could be estimated from a functional form derived from publicly available prices for Australian generators coupled with several spot prices for US-made generators.

As discussed in OECD Nuclear Energy Agency (2010b, 2011), it is nearly impossible to exactly determine the functional form for the discarding cost, $\hat{z}_a(f_a)$, since these costs are not separately reported by the processors and, due to security issues involving HEU waste, discarding costs at the processor are often paid for by the government. In general, however, the functional form of $\hat{z}_a(f_a)$ should be consistent with $\hat{c}_a(f_a)$.

Moreover, the discarding cost of LEU processing (link 24) should be about twice the discarding costs for the HEU processing (link 3); the costs of discarding during elucitation are higher in the USA (link 18) as compared to other countries (links 19 and 20), and the discarding costs for all generator manufacturers are the same.

3.4 A Case Study

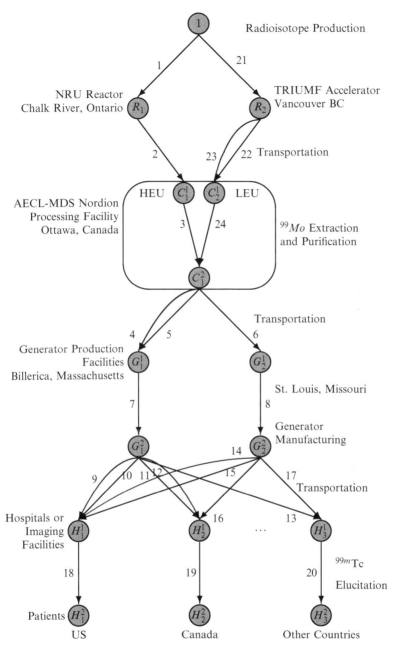

Fig. 3.3 The case study medical nuclear supply chain topology for ^{99}Mo from Canada to the United States, Canada, and other countries

The risk functions for transportation links were estimated based upon the overall accident rate per kilometer for aircraft and trucks carrying nuclear material as reported in Resnikoff (1992). These were converted to a per curie basis using an average distance traveled and the approximate number of curies transported per week. The functions for risk for shipment of bulk ^{99}Mo and irradiated targets were increased by 1 and 2 orders of magnitude, respectively, to account for the increased severity of an accident during this stage. The risk during generator elucitation was assumed to be similar to air transport of a generator ($\approx 2 \times 10^{-6} f_x$). Risks during generator production, bulk ^{99}Mo extraction, and irradiation were assumed to be 1, 2, and 3 orders of magnitude greater, respectively.

We assumed three demand points corresponding, respectively, to the collective demands in the USA, in Canada, and in other countries (such as Mexico and the Caribbean Islands). We are using three demand points, as approximations, in order to be able to report the input and output data for transparency purposes. The demands were as follows: $d_1 = 3,600$, $d_2 = 1,800$, and $d_3 = 1,000$, and these denote the demands, in curies, per week. These values were obtained by using the daily number of procedures in the USA and extrapolating for the others. The units for the link flows are also curies. We initialized the algorithm by equally distributing the O/D demand among its paths to determine the initial path flow pattern; the Lagrange multipliers were initialized at zero.

Scenario 1. Above we have described the fundamental setting of our case study. In Scenario 1, the weight ω was set equal to 1. Note that this weight can also represent a conversion factor of risk to cost.

The arc multipliers, the total operational cost functions, the total discarding cost functions, the total risk functions, and the optimal link flow and Lagrange multiplier solutions are reported in Table 3.1.

As can be seen from the results in Table 3.1, none of the links were operated at full capacity, and hence, all the Lagrange multipliers were equal to 0.00. Note also that the transportation links 10, 12, 15, 16, and 22 have zero flow. Hence, not only does cost efficiency play a role but also so does the perishability of the product. Transport by truck may be cheaper, but it takes longer, and hence, there may be less product available once the shipment is delivered. Note that the LEU processing link 24 had positive flow, and air transport (given the distance) was used (see flow on link 23) to deliver the LEU targets to the processing facility. This makes sense, given the time-sensitivity and perishability of molybdenum and the distance between the Vancouver accelerator and the processing facility in Ottawa. Also, it is interesting to observe, as speculated by Kramer (2011), that the new LEU production facility (cf. link 21) is expected to produce about 30% of the needs, and this was also the result obtained in our computation. The HEU reactor in Chalk River (cf. link 1) produced 9,720.70 curies, whereas the Vancouver one produced 5,867.24 curies (cf. link 21).

The value of the objective function [cf. (3.18)] was 2,096,149,888.00, whereas the total risk [cf. (3.16)] was 4,060.81. It is also important to emphasize that the

3.4 A Case Study

Table 3.1 Link multipliers, link total operational cost, total discarding cost, total risk functions, link capacity, and optimal link flow and Lagrange multiplier solution for the Scenario 1

Link a	α_a	$\hat{c}_a(f_a)$	$\hat{z}_a(f_a)$	$\hat{r}_a(f_a)$	$\bar{u}_a$	f_a^*	λ_a^*
1	1.00	$2f_1^2+25.6f_1$	0.00	$2.00 \times 10^{-2} f_1$	33,353	9,720.20	0.00
2	.969	$f_2^2+5f_2$	0.00	$3.18 \times 10^{-1} f_2$	large	9,720.20	0.00
3	.706	$5f_3^2+192f_3$	$5f_3^2+80f_3$	$2.00 \times 10^{-3} f_3$	32,154	9,419.35	0.00
4	.920	$2f_4^2+4f_4$	0.00	$1.59 \times 10^{-1} f_4$	large	3,045.90	0.00
5	.901	$f_5^2+f_5$	0.00	$2.16 \times 10^{-3} f_5$	large	2,327.34	0.00
6	.915	$f_6^2+2f_6$	0.00	$6.9 \times 10^{-3} f_6$	large	4,934.45	0.00
7	.804	$f_7^2+166f_7$	$2f_7^2+7f_7$	$2.00 \times 10^{-4} f_7$	19,981	4,899.16	0.00
8	.804	$f_8^2+166f_8$	$2f_7^2+7f_7$	$2.00 \times 10^{-4} f_8$	19,981	4,515.02	0.00
9	.883	$2f_9^2+4f_9$	0.00	$2.00 \times 10^{-4} f_9$	large	1,623.06	0.00
10	.779	$f_{10}^2+1f_{10}$	0.00	$1.47 \times 10^{-2} f_{10}$	large	0.00	0.00
11	.883	$2f_{11}^2+4f_{11}$	0.00	$2.00 \times 10^{-4} f_{11}$	large	2,038.51	0.00
12	.688	$f_{12}^2+2f_{12}$	0.00	$1.47 \times 10^{-2} f_{12}$	large	0.00	0.00
13	.688	$2.5f_{13}^2+2f_{13}$	0.00	$1.98 \times 10^{-4} f_{13}$	large	277.36	0.00
14	.883	$2f_{14}^2+2f_{14}$	0.00	$7.33 \times 10^{-3} f_{14}$	large	2,453.95	0.00
15	.779	$f_{15}^2+7f_{15}$	0.00	$1.00 \times 10^{-4} f_{15}$	large	0.00	0.00
16	.688	$2f_{16}^2+4f_{16}$	0.00	$1.00 \times 10^{-4} f_{16}$	large	0.00	0.00
17	.688	$2f_{17}^2+6f_{17}$	0.00	$1.98 \times 10^{-5} f_{17}$	large	1,176.13	0.00
18	1.00	$2f_{18}^2+800f_{18}$	$4f_{18}^2+80f_{18}$	$2.00 \times 10^{-5} f_{18}$	5,000	3,600.00	0.00
19	1.00	$f_{19}^2+600f_{19}$	$1f_{19}^2+60f_{19}$	$2.00 \times 10^{-5} f_{19}$	3,000	1,800.00	0.00
20	1.00	$f_{20}^2+300f_{20}$	$1f_{20}^2+30f_{20}$	$2.00 \times 10^{-5} f_{20}$	2,000	1,000.00	0.00
21	1.00	$4f_{21}^2+50f_{21}$	0.00	$2.00 \times 10^{-2} f_{21}$	10,006	5,867.24	0.00
22	.436	$6f_{22}^2+6f_{22}$	0.00	$1.04 \times 10^{1} f_{22}$	large	0.00	0.00
23	.883	$3f_{23}^2+21f_{23}$	0.00	$1.44 \times 10^{-2} f_{23}$	large	5,867.24	0.00
24	.706	$5f_{24}^2+192f_{24}$	$10f_{24}^2+160f_{24}$	$2.00 \times 10^{-3} f_{24}$	10,006	5,180.77	0.00

demand was met at the demand points but, given the perishability of the radioisotope, many more curies had to be produced than were demanded, with $f_1^* + f_{21}^* = 15,587.44$ curies produced per week for the total demand of $d_1 + d_2 + d_3 = 6,400$ curies.

Scenario 2. We then proceeded to construct Scenario 2 of our case study (see the data in Table 3.2). We raised the evaluation of the weight ω from 1 to 1,000. The remainder of the input data were as in Table 3.1. The computed optimal solution for Scenario 2 is also reported in Table 3.2.

As can be seen from Table 3.2, the same links as in Scenario 1 had zero flows. The value of the objective function was now 2,100,204,416.00, and the total risk was now 4,055.70. The risk was reduced, as expected, since ω was increased. Note, for example, that the link flows shifted from links with higher total risk to those with lower total risk, such as the shift from link 1 to link 21. However, the total value of the objective function increased due to the higher value of ω. The total number of curies produced per week was now $f_1^* + f_{21}^* = 15,589.10$ with the demand as in Scenario 1, that is, 6,400.

Table 3.2 Link multipliers, link total operational cost, total discarding cost, total risk functions, link capacity, and optimal link flow and Lagrange multiplier solution for Scenario 2

Link a	α_a	$\hat{c}_a(f_a)$	$\hat{z}_a(f_a)$	$\hat{r}_a(f_a)$	$\bar{u}_a$	f_a^*	λ_a^*
1	1.00	$2f_1^2 + 25.6f_1$	0.00	$2.00 \times 10^{-2} f_1$	33,353	9,716.86	0.00
2	.969	$f_2^2 + 5f_2$	0.00	$3.18 \times 10^{-1} f_2$	large	9,716.86	0.00
3	.706	$5f_3^2 + 192f_3$	$5f_3^2 + 80f_3$	$2.00 \times 10^{-3} f_3$	32,154	9,415.64	0.00
4	.920	$2f_4^2 + 4f_4$	0.00	$1.59 \times 10^{-1} f_4$	large	3,020.34	0.00
5	.901	$f_5^2 + f_5$	0.00	$2.16 \times 10^{-3} f_5$	large	2,350.27	0.00
6	.915	$f_6^2 + 2f_6$	0.00	$6.9 \times 10^{-3} f_6$	large	4,937.57	0.00
7	.804	$f_7^2 + 166f_7$	$2f_7^2 + 7f_7$	$2.00 \times 10^{-4} f_7$	19,981	4,896.31	0.00
8	.804	$f_8^2 + 166f_8$	$2f_7^2 + 7f_7$	$2.00 \times 10^{-4} f_8$	19,981	4,517.88	0.00
9	.883	$2f_9^2 + 4f_9$	0.00	$2.00 \times 10^{-4} f_9$	large	1,622.27	0.00
10	.779	$f_{10}^2 + 1f_{10}$	0.00	$1.47 \times 10^{-2} f_{10}$	large	0.00	0.00
11	.883	$2f_{11}^2 + 4f_{11}$	0.00	$2.00 \times 10^{-4} f_{11}$	large	2038.51	0.00
12	.688	$f_{12}^2 + 2f_{12}$	0.00	$1.47 \times 10^{-2} f_{12}$	large	0.00	0.00
13	.688	$2.5f_{13}^2 + 2f_{13}$	0.00	$1.98 \times 10^{-4} f_{13}$	large	275.86	0.00
14	.883	$2f_{14}^2 + 2f_{14}$	0.00	$7.33 \times 10^{-3} f_{14}$	large	2,454.74	0.00
15	.779	$f_{15}^2 + 7f_{15}$	0.00	$1.00 \times 10^{-4} f_{15}$	large	0.00	0.00
16	.688	$2f_{16}^2 + 4f_{16}$	0.00	$1.00 \times 10^{-4} f_{16}$	large	0.00	0.00
17	.688	$2f_{17}^2 + 6f_{17}$	0.00	$1.98 \times 10^{-5} f_{17}$	large	1,177.63	0.00
18	1.00	$2f_{18}^2 + 800f_{18}$	$4f_{18}^2 + 80f_{18}$	$2.00 \times 10^{-5} f_{18}$	5,000	3,600.00	0.00
19	1.00	$f_{19}^2 + 600f_{19}$	$1f_{19}^2 + 60f_{19}$	$2.00 \times 10^{-5} f_{19}$	3,000	1,800.00	0.00
20	1.00	$f_{20}^2 + 300f_{20}$	$1f_{20}^2 + 30f_{20}$	$2.00 \times 10^{-5} f_{20}$	2,000	1,000.00	0.00
21	1.00	$4f_{21}^2 + 50f_{21}$	0.00	$2.00 \times 10^{-2} f_{21}$	10,006	5,872.24	0.00
22	.436	$6f_{22}^2 + 6f_{22}$	0.00	$1.04 \times 10^{1} f_{22}$	large	0.00	0.00
23	.883	$3f_{23}^2 + 21f_{23}$	0.00	$1.44 \times 10^{-2} f_{23}$	large	5,872.24	0.00
24	.706	$5f_{24}^2 + 192f_{24}$	$10f_{24}^2 + 160f_{24}$	$2.00 \times 10^{-3} f_{24}$	10,006	5,185.19	0.00

Scenario 3. We then constructed Scenario 3. The data were as for Scenario 2, but now we considered a medical nuclear supply chain disruption in the form of a capacity reduction at the NRU reactor with its capacity being reduced to 9,000. The complete input and output data are reported in Table 3.3. Note that, since the capacity at link 1 was reached, there was a positive associated Lagrange multiplier on that link.

The value of the objective function was now 2,117,958,400.00, and the total risk was 3,865.07. Interestingly, this was the lowest value obtained in the three scenarios. For this example, the total risk decreased even more significantly than that observed for Scenario 2 relative to Scenario 1. As the capacity associated with link 1 decreased, in order to satisfy the demands and minimize the costs (and losses), the link flows shifted to links with higher capacities and larger arc multipliers, such as the shift from link 1 to 21 and the shift from link 5 to link 4, although link 4 was the one with higher total cost. The total amount of curies that was needed to be produced was now $f_1^* + f_{21}^* = 15,650.96$ with the demand at $6,400$.

3.5 Summary and Conclusions

Table 3.3 Link multipliers, link total operational cost, total discarding cost, total risk functions, link capacity, and optimal link flow and Lagrange multiplier solution for Scenario 3

Link a	α_a	$\hat{c}_a(f_a)$	$\hat{z}_a(f_a)$	$\hat{r}_a(f_a)$	$\bar{u}_a$	f_a^*	λ_a^*
1	1.00	$2f_1^2 + 25.6f_1$	0.00	$2.00 \times 10^{-2} f_1$	9,000.00	9,000.00	4,9667.88
2	.969	$f_2^2 + 5f_2$	0.00	$3.18 \times 10^{-1} f_2$	large	9,000.00	0.00
3	.706	$5f_3^2 + 192f_3$	$5f_3^2 + 80f_3$	$2.00 \times 10^{-3} f_3$	32,154	8,722.89	0.00
4	.920	$2f_4^2 + 4f_4$	0.00	$1.59 \times 10^{-1} f_4$	large	3,173.72	0.00
5	.901	$f_5^2 + f_5$	0.00	$2.16 \times 10^{-3} f_5$	large	2,168.41	0.00
6	.915	$f_6^2 + 2f_6$	0.00	$6.9 \times 10^{-3} f_6$	large	4,962.43	0.00
7	.804	$f_7^2 + 166f_7$	$2f_7^2 + 7f_7$	$2.00 \times 10^{-4} f_7$	19,981	4,873.56	0.00
8	.804	$f_8^2 + 166f_8$	$2f_7^2 + 7f_7$	$2.00 \times 10^{-4} f_8$	19,981	4,540.62	0.00
9	.883	$2f_9^2 + 4f_9$	0.00	$2.00 \times 10^{-4} f_9$	large	1,612.58	0.00
10	.779	$f_{10}^2 + 1f_{10}$	0.00	$1.47 \times 10^{-2} f_{10}$	large	0.00	0.00
11	.883	$2f_{11}^2 + 4f_{11}$	0.00	$2.00 \times 10^{-4} f_{11}$	large	2038.51	0.00
12	.688	$f_{12}^2 + 2f_{12}$	0.00	$1.47 \times 10^{-2} f_{12}$	large	0.00	0.00
13	.688	$2.5f_{13}^2 + 2f_{13}$	0.00	$1.98 \times 10^{-4} f_{13}$	large	267.26	0.00
14	.883	$2f_{14}^2 + 2f_{14}$	0.00	$7.33 \times 10^{-3} f_{14}$	large	2,464.43	0.00
15	.779	$f_{15}^2 + 7f_{15}$	0.00	$1.00 \times 10^{-4} f_{15}$	large	0.00	0.00
16	.688	$2f_{16}^2 + 4f_{16}$	0.00	$1.00 \times 10^{-4} f_{16}$	large	0.00	0.00
17	.688	$2f_{17}^2 + 6f_{17}$	0.00	$1.98 \times 10^{-5} f_{17}$	large	1,186.23	0.00
18	1.00	$2f_{18}^2 + 800f_{18}$	$4f_{18}^2 + 80f_{18}$	$2.00 \times 10^{-5} f_{18}$	5,000	3,600.00	0.00
19	1.00	$f_{19}^2 + 600f_{19}$	$1f_{19}^2 + 60f_{19}$	$2.00 \times 10^{-5} f_{19}$	3,000	1,800.00	0.00
20	1.00	$f_{20}^2 + 300f_{20}$	$1f_{20}^2 + 30f_{20}$	$2.00 \times 10^{-5} f_{20}$	2,000	1,000.00	0.00
21	1.00	$4f_{21}^2 + 50f_{21}$	0.00	$2.00 \times 10^{-2} f_{21}$	10,006	6,650.96	0.00
22	.436	$6f_{22}^2 + 6f_{22}$	0.00	$1.04 \times 10^{1} f_{22}$	large	0.00	0.00
23	.883	$3f_{23}^2 + 21f_{23}$	0.00	$1.44 \times 10^{-2} f_{23}$	large	6,650.96	0.00
24	.706	$5f_{24}^2 + 192f_{24}$	$10f_{24}^2 + 160f_{24}$	$2.00 \times 10^{-3} f_{24}$	10,006	5,872.80	0.00

Interestingly, in Scenarios 1, 2, and 3, links 10, 12, 15, 16, and 22 all had zero flows at the optimal solution. If we eliminate such links from Fig. 3.3, what may be viewed as the *final* supply chain network topology in which all the links have positive flows would be as in Fig. 3.4.

We now also provide the computed optimal path flow solution for Scenario 3. Please refer to Fig. 3.2 for the link numerical identification. For O/D pair $w_1 = (1, H_1^2)$, there are 18 available paths; for O/D pair $w_2 = (1, H_2^2)$, there are 15 available paths; and for O/D pair $w_3 = (1, H_3^2)$, the number of available paths is 9. In Table 3.4, we report the flows for those paths that had positive flows at the computed optimal solution—all other paths had zero flows and, hence, were not used.

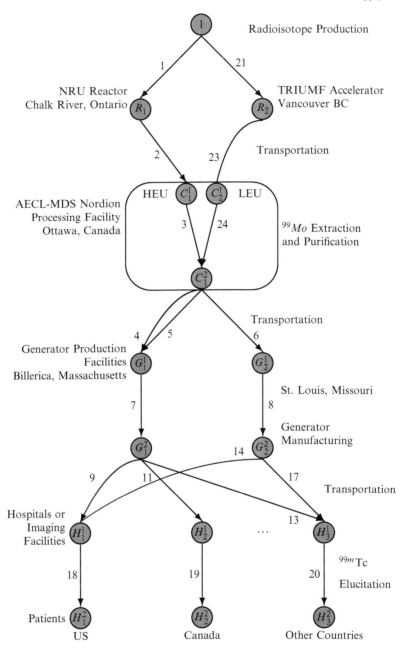

Fig. 3.4 The case study *final* medical nuclear supply chain topology for ^{99}Mo from Canada to the United States, Canada, and other countries with only links with positive flows displayed for Scenarios 1, 2, and 3

Table 3.4 Optimal path flow pattern for Scenario 3 with paths with positive flows reported

	Path definition	Path flow
O/D pair $w_1 = (1, H_1^2)$	$p_1 = (1,2,3,4,7,9,18)$	$x_{p_1}^* = 1,142.06$
	$p_2 = (1,2,3,5,7,9,18)$	$x_{p_2}^* = 861.74$
	$p_3 = (1,2,3,6,8,14,18)$	$x_{p_3}^* = 2,799.40$
	$p_4 = (21,23,24,4,7,9,18)$	$x_{p_4}^* = 804.59$
	$p_5 = (21,23,24,5,7,9,18)$	$x_{p_5}^* = 523.94$
	$p_6 = (21,23,24,6,8,14,18)$	$x_{p_6}^* = 2,301.64$
O/D pair $w_2 = (1, H_2^2)$	$p_1 = (1,2,3,4,7,11,19)$	$x_{p_1}^* = 1,380.63$
	$p_2 = (1,2,3,5,7,11,19)$	$x_{p_2}^* = 1,100.50$
	$p_3 = (21,23,24,4,7,11,19)$	$x_{p_3}^* = 1,014.54$
	$p_4 = (21,23,24,5,7,11,19)$	$x_{p_4}^* = 723.38$
O/D pair $w_3 = (1, H_3^2)$	$p_1 = (1,2,3,4,7,13,20)$	$x_{p_1}^* = 286.51$
	$p_2 = (1,2,3,5,7,13,20)$	$x_{p_2}^* = 11.41$
	$p_3 = (1,2,3,6,8,17,20)$	$x_{p_3}^* = 1,419.69$
	$p_4 = (21,23,24,4,7,13,20)$	$x_{p_4}^* = 189.07$
	$p_5 = (21,23,24,5,7,13,20)$	$x_{p_5}^* = 65.18$
	$p_6 = (21,23,24,6,8,17,20)$	$x_{p_6}^* = 1,028.63$

3.5 Summary and Conclusions

In this chapter, we presented a medical nuclear supply chain optimization model that incorporates the time-dependent and perishable nature of radioisotopes as well as the hazardous aspects which affect not only the transportation modes that can be used but also waste and risk management issues. The model reflects the realities of the technology in this important healthcare sector that is transitioning globally from highly enriched uranium targets to low enriched uranium ones. The model is a generalized network model and includes multicriteria decision-making so that the organization can minimize both total operating and waste management costs as well as the risk associated with the various supply chain network activities of processing, generator production, transportation, and ultimate usage in medical procedures at hospitals and other medical facilities. Each link has associated with it an arc multiplier that is constructed based on physics principles of radioactive decay of the radioisotope—in this case, molybdenum. Hence, the model traces the amount of the radioisotope that is left as a particular pathway of the supply chain is traversed. Capacities on the links reflect the realities of the production, processing, generation, and transportation activities.

The formulation of the model and the qualitative analysis exploit variational inequality theory since it yields a very elegant procedure for computational purposes. Moreover, it provides us with the foundation to explore other scenarios as the technology landscape continues to evolve and to bring other participants into medical nuclear production. A numerical case study based on North America, with the focus of an existing HEU reactor and an LEU accelerator that is expected to come online soon, reveals the generality and practicality of our framework.

The focus in this chapter, as in Chap. 2, was on system optimization since the emphasis was on economic cost recovery and transparency. An interesting question for future research would be the investigation of various types of possible competition associated with, for example, transportation service providers in the medical nuclear supply chain arena as well as competition among the generator manufacturing facilities. To develop such models, one may utilize the concepts associated with competitive supply chain network problems governed by Nash equilibria. We explore aspects of such issues in other applications in Part III of this book.

3.6 Sources and Notes

This chapter is based on the paper by Nagurney et al. (2012). Another related paper, which focuses on the design of medical nuclear supply chains, is by Nagurney and Nagurney (2012). In this chapter, we also report path flow solutions and provide additional discussion over and above that in the former study. The design of supply chain networks, under optimization, has been formulated as a variational inequality problem in a general setting by Nagurney (2010). In that model, the link capacities as well as the path and link flows are variables. In addition, it is worth noting that one can also formulate supply chain network optimization models that include frequencies of operating the various supply chain network activities of production, shipment, etc., as was done in the paper Nagurney (2012) that concentrated on the sustainable supply chains for sustainable cities.

Recent work that is relevant to supply chain optimization of other hazardous materials can be categorized as follows: capacity planning, facility location, or routing and scheduling (see Erkut et al. 2007). Most of the studies apply integer programming, stochastic programming, or simulation as the methodology of choice. In addition, although risk minimization has been taken into account in the hazmat transportation literature in addition to cost minimization (see, e.g., Batta and Chiu 1988 and Erkut and Gzara 2008), our approach is the first to consider risk for medical nuclear supply chain optimization.

It is important to note that traditional risk functions estimate the risk as the product of the probability or the conditional probability of an accident happening and its consequences (cf. Sherali et al. 1997), in which the total probability of an accident happening on a road segment is the product of the unit probability, road length and travel time (in the case of land-based transportation); the consequence of an accident is estimated by the number of shipments, the population exposure, and the area of the impact zone. Thus, it is reasonable to assume that, in the case of hazmat transportation, the risk functions depend on the mode and flows, as in this chapter.

In our case study, we showed the effects of varying the factor of risk aversion as well as a disruption in the production capacity on the optimal production, processing, and other supply chain activities of the radioisotope molybdenum, in order that the supply met the demand. Our framework vividly illustrates how the perishability of a product, which, in this chapter was a radioisotope, which is subject to radioactive decay over time, affects the total costs and the risks.

The models in Part II of this book are multicriteria system-optimization models, focused on healthcare applications. Qiang and Nagurney (2012) also used a system-optimization framework for their supply chain network model for critical needs products, which captures disruptions in capacities associated with the various supply chain activities of production, transportation, and storage, as well as those associated with the demands for the product at the various demand points. The authors proposed two distinct supply chain network performance measures for such products, which are needed in healthcare and also in the context of emergencies and disasters. The first measure considers disruptions in the link capacities but assumes that the demands for the product can be met. The second measure captures the unsatisfied demand. They then proposed a bi-criteria supply chain network performance measure and used it for the evaluation of distinct supply chain networks.

References

Alp E (1995) Risk-based transportation planning practice: Overall methodology and a case example. Inform Syst Oper Res 33(1):4–19

American Institute of Physics (1972) American Institute of Physics Handbook, 3rd edn. McGraw Hill, New York, 8-6-8-91

Batta R, Chiu SS (1988) Optimal obnoxious paths on a network: transportation of hazardous materials. Oper Res 36(1):84–92

Berger SA, Goldsmith W, Lewis ER (eds) (2004) Introduction to bioengineering. Oxford University Press, Oxford

Dafermos SC, Sparrow FT (1969) The traffic assignment problem for a general network. J Res Natl Bur Stand 73B:91–118

de Lange F (2010) Covidien's role in the supply chain of Molybdenum-99 and Technetium-99m generators. Tijdschrift voor Nucleaire Geneeskunde 32:593–596

Erkut E, Gzara F (2008) Solving the hazmat transport network design problem. Comput Oper Res 35(7):2234–2247

Erkut E, Tjandra SA, Verter V (2007) Hazardous materials transportation. Handbooks Oper Res Manag Sci 14:539–621

Kahn LH (2008) The potential dangers in medical isotope production. Bulletin of the Atomic Scientists, March 16. Available online at: `http://thebulletin.org/node/163`

Kochanek KD, Xu J, Murphy SL, Minino AM, Kung H-C (2011) Deaths: preliminary data for 2009, National Vital Statistics Reports, vol 59, No. 4, March 16

Korpelevich GM (1977) The extragradient method for finding saddle points and other problems. Matekon 13:35–49

Kramer D (2011) Drive to end civilian use of HEU collides with medical isotope production. Phys Today 64(2):17–19

Lantheus Medical Imaging, Inc. (2009) Lantheus Medical Imaging takes proactive steps to mitigate impact of global molybdenun-99 supply chain operations on operations. Press Release, May 20. Available online at: http://www.lantheus.com/News-Press-2009-0520.html

Nagurney A (2010) Optimal supply chain network design and redesign at minimal total cost and with demand satisfaction. Int J Prod Econ 128(1):200–208

Nagurney A (2012) Design of sustainable supply chains for sustainable cities. Isenberg School of Management, University of Massachusetts, Amherst, MA

Nagurney A, Nagurney LS (2012) Medical nuclear supply chain design: A tractable network model and computational approach. Int J Prod Econ 140(2):865–874

Nagurney A, Nagurney LS, Li D (2012) Securing the sustainability of global medical nuclear supply chains through economic cost recovery, risk management, and optimization. Int J Sustain Transport (to appear)

Nagurney A, Qiang Q (2009) Fragile networks: identifying vulnerabilities and synergies in an uncertain world. Wiley, Hoboken, NJ

National Research Council (2009) Medical isotope production without highly enriched uranium. The National Academy of Sciences, Washington, DC

OECD Nuclear Energy Agency (2010a) The supply of medical radioisotopes: Interim report of the OECD/NEA high-level group on security of supply of medical radioisotopes

OECD Nuclear Energy Agency (2010b) The supply of medical radioisotopes: An economic study of the molybdenum-99 supply chain NEA No. 6867

OECD Nuclear Energy Agency (2011) The supply of medical radioisotopes: The path to reliability NEA No. 6985

Ponsard B (2010) Mo-99 supply issues: Report and lessons learned, presented at the 14th International Topical Meeting on the Research Reactor Fuel Management (RRFM 2010), Marrakech, Morrocco, 21–25 March, published by the European Nuclear Society, ENS RRFM 2010 Transactions

Qiang Q, Nagurney A (2012) A bi-criteria measure to assess supply chain network performance for critical needs under capacity and demand disruptions. Transport Res A 46(5):801–812

Resnikoff M (1992) Study of transportation accident severity. Radioactive Waste Management Associates for the Nevada Nuclear Waste Project Office, February

Seeverens HJJ (2010) The economics of the molybdenum-99/Technetium-99m supply chain. Tijdschrift voor Nucleaire Geneeskunde 32:604–608

Sherali HD, Brizendine LD, Glickman TS, Subramianian S (1997) Low probability-high consequence considerations in routing hazardous material shipments. Transport Sci 31(3):237–251

World Nuclear Association (2012) Radioisotopes in medicine. January 26. Available online at: http://www.world-nuclear.org/info/inf55.html

Part III
Supply Chain Network Game Theory Models and Applications

Chapter 4
Food Supply Chains

Abstract In this chapter, we construct a food supply chain network model under oligopolistic competition and perishability, with a focus on fresh produce. The model handles food spoilage through arc multipliers, with the inclusion of the discarding costs associated with disposal. We allow for product differentiation due to such relevant issues as product freshness and food safety and include alternative technologies associated with various supply chain activities. A case study, in which we analyze different scenarios prior to, during, and post a foodborne disease outbreak, based on the cantaloupe market, illustrates the modeling and computational framework.

4.1 Motivation and Overview

This chapter begins Part III of the book, which focuses on game theoretical supply chain networks for perishable products. Unlike the single-organization optimization models in Part II, the models in this part are multiple decision-maker competitive equilibrium models. These supply chain network problems are formulated under product/brand differentiation to capture the reality of real-world competition and to provide the appropriate analytics.

Food supply chains have several characteristic differences from other product supply chains and, hence, merit individual consideration and treatment. The fundamental difference between food supply chains and other supply chains is the continuous change in the quality of food products throughout the entire supply chain. This is especially important for fresh produce supply chains where increasing attention is being placed on both product freshness and safety. Moreover, growing demand for such products is supported by statistics from the United States Department of Agriculture (USDA 2011), which suggests that the consumption of fresh vegetables has increased at a much faster pace than the demand for traditional crops.

Today's food supply chains are complex global networks that create pathways from farms to consumers, involving production, processing, distribution, and even the disposal of food (see Ahumada and Villalobos 2009). Consumers' expectation of year-around availability of fresh food products has encouraged the globalization of food markets. With growing global competition and the greater distances between food production and consumption locations, there is increasing pressure for the integration of food production and distribution. This results in new challenges for food supply chain modeling, analysis, and solutions.

Given the thin profit margins in food industries, product differentiation strategies are increasingly being used with product freshness considered as one of the major differentiating factors. Retailers now realize that food freshness can be a competitive advantage (Lütke Entrup 2005; see also Aiello et al. 2012).

Moreover, the high perishability of food products results in immense food waste/loss, further stressing food supply chains. While some food waste and loss are inevitable on food supply chain network links, it is estimated that approximately one third of the global food production is wasted or lost annually (Gustavsson et al. 2011). Food products often need special handling, transportation, and storage technologies. Furthermore, the quality of fresh food products decreases with time, despite the use of the most advanced processing, handling, and shipping methods.

In this chapter, we formulate, analyze, and solve food supply chain network problems, under product differentiation and competition, including food deterioration/spoilage. We focus on fresh produce items, such as vegetables and fruits, that require simple or limited processing and whose lifespans can be measured in days. The fresh produce supply chain network oligopoly model developed in this chapter is distinct from other studies on perishable food products in that:

- It captures the deterioration of fresh food along the entire supply chain from a network perspective.
- It handles the exponential time decay through the introduction of arc multipliers.
- It captures oligopolistic competition with product differentiation.
- It includes the disposal of the spoiled food products, along with the associated costs.
- It allows for the assessment of alternative technologies for each supply chain activity.

This chapter is organized as follows. In Sect. 4.2, we construct the food supply chain network oligopoly model and derive variational inequality formulations. We also provide some qualitative properties. In Sect. 4.3, we present the computational algorithm, which we then apply to a case study focused on fresh produce in Sect. 4.4. We summarize our results and present our conclusions in Sect. 4.5. We conclude with a Sources and Notes section.

4.2 The Food Supply Chain Network Oligopoly Model

In this section, we consider a food supply chain network with a finite number of I food firms, with a typical firm denoted by i. The food supply chain network activities include production, processing, storage, distribution, and the disposal of the food

4.2 The Food Supply Chain Network Oligopoly Model

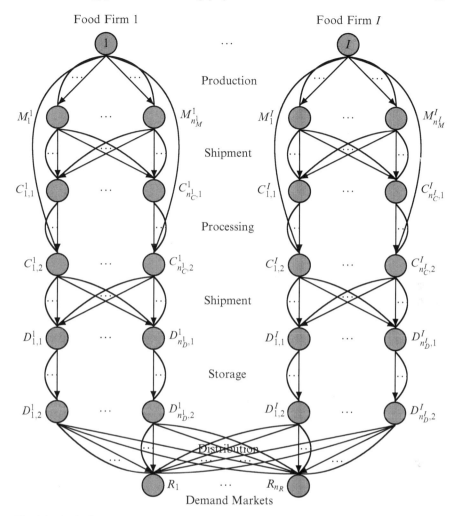

Fig. 4.1 The fresh produce supply chain network topology

products. The firms will typically be vertically integrated, which, as a strategy, has become increasingly important as food systems become more consumer-driven (see Bhuyan 2005).

The individual food firms compete noncooperatively in an oligopolistic manner. We allow for product differentiation by consumers at the demand markets, due to, for example, product freshness and food safety concerns that may be associated with a particular firm. This means that the fresh food products of a given type are not necessarily homogeneous.

Food firms are represented by supply chain networks of their economic activities as depicted in Fig. 4.1. Each firm seeks to determine its optimal product flows.

Each food firm i possesses n_M^i production facilities, n_C^i processors, and n_D^i distribution centers to satisfy the demands at n_R demand markets. Let $\mathscr{G} = [N, L]$ denote the graph consisting of the set of nodes N and the set of links L in Fig. 4.1, where $L \equiv \cup_{i=1,\ldots,I} L^i$ and L^i denotes the set of directed links corresponding to the sequence of activities associated with firm i.

The first set of links connecting the top two tiers of nodes corresponds to the food production at each of the production units of firm i. Production may involve seasonal operations such as soil agitation, sowing, pest control, nutrient and water management, and harvesting. The multiple possible links connecting each top tier node i with its production facilities, $M_1^i, \ldots, M_{n_M^i}^i$, represent alternative production technologies that may be associated with a given facility.

The second set of links from the production facility nodes is connected to the processors of each firm i which are denoted by $C_{1,1}^i, \ldots, C_{n_C^i,1}^i$. These links correspond to the shipment links between the production units and the processors. The multiple shipment links denote different possible modes of transportation, characterized by varying time durations and environmental conditions.

The third set of links connecting nodes $C_{1,1}^i, \ldots, C_{n_C^i,1}^i$ to $C_{1,2}^i, \ldots, C_{n_C^i,2}^i$ denotes the processing of fresh produce. The major food processing activities include cleaning, sorting, labeling, and simple packaging. Different processing activities and technologies may result in different levels of quality degradation.

The next set of nodes represents the distribution centers. Thus, the fourth set of links connecting the processor nodes to the distribution centers is the set of shipment links. Such distribution nodes associated with firm i are denoted by $D_{1,1}^i, \ldots, D_{n_D^i,1}^i$. There are also multiple shipment links in order to capture different modes of transportation.

The fifth set of links, which connects nodes $D_{1,1}^i, \ldots, D_{n_D^i,1}^i$ to $D_{1,2}^i, \ldots, D_{n_D^i,2}^i$, are the storage links. Since fresh produce items may require alternative storage conditions, we represent these alternative conditions through multiple links at this tier.

The last set of links connecting the two bottom tiers of the supply chain network corresponds to the distribution links over which the fresh produce items are shipped from the distribution centers to the demand markets. Here we also allow for multiple modes of transportation.

In addition, the curved links in Fig. 4.1 joining the top-tiered nodes i with the processors, which are denoted by $C_{1,2}^i, \ldots, C_{n_C^i,2}^i$, capture the possibility of on-site production and processing.

Most of the fresh produce items reach their peak quality at the time of production and then deteriorate over time (Blackburn and Scudder 2009). Microbiological decay is one of the major causes of food quality degradation, especially for fresh produce (Fu and Labuza 1993). Therefore, food deterioration usually follows a first-order equation with exponential time decay (see Labuza 1982, Nahmias 1982, Tijskens and Polderdijk 1996, Blackburn and Scudder 2009, Nga 2010, and Rong 2011). Exponential time decay is a special case of random lifetime perishability, which means that, while the time to spoilage of an individual item is uncertain, the percentage of products that are spoiled at a given time can be predicted (Nahmias 1982; see also Van Zyl 1964). It also has been recognized that the decay constant is highly dependent on the temperatures and other environmental conditions. Food supply chains can be grouped into three types based on various temperature requirements: frozen, chilled, and ambient. The normal temperature of the frozen chain is $-18°C$, while temperatures range from $0°C$ for fresh fish to $15°C$ for potatoes and bananas in the chilled chain. There is no required temperature control in an ambient chain.

In the existing literature on perishability, exponential time decay has been utilized to describe either the decrease in quantity or the degradation in quality. The decrease in quantity, which has been discussed in studies on perishable inventory, represents the number of units of decayed products such as vegetables and fruits. On the other hand, the degradation in quality emphasizes that all the products deteriorate at the same rate simultaneously, which is more relevant to meat, dairy, and bakery products. Since our model focuses on fresh produce items, we adopt exponential time decay to capture the discarding of spoiled products associated with all the postproduction activities. We will highlight how the model can be modified/adapted to handle degradation in quality.

4.2.1 The Underlying Chemistry

Food products deteriorate over time even under optimal conditions. We assume that the temperature and other environmental conditions associated with each postproduction activity/link are known and fixed. Following Nahmias (1982), we assume that each unit has a probability of $e^{-\lambda t}$ to survive another t units of time, where λ is a positive parameter known as the decay constant. Let N_0 denote the quantity at the beginning of the time interval (link). Then, the expected quantity surviving at the end of the time interval (specific link), denoted by $N(t)$, can be expressed as

$$N(t) = N_0 e^{-\lambda t}. \tag{4.1}$$

As in Chaps. 2 and 3, we assign a multiplier to each link in the supply chain network. Here, as in Chap. 3, we are interested in capturing the decay in the number of units. Hence, we can represent the arc multiplier α_a for a postproduction link a as

$$\alpha_a = e^{-\lambda_a t_a}, \tag{4.2}$$

where λ_a and t_a are the decay constant and the time duration associated with the link a. Both λ_a and t_a are given and fixed. We assume that the value of α_a for a production link is equal to one.

In rare cases, food deterioration follows the zero-order reactions with linear decay (Tijskens and Polderdijk 1996 and Rong 2011). Then, $\alpha_a = 1 - \lambda_a t_a$ for a postproduction link.

For definiteness and since we are focusing on the fresh produce application, we express the flows, the multipliers, and their relationships in this context.

As noted in Chaps. 2 and 3, f_a denotes the (initial) flow of product on link a, and f'_a denotes the final flow on link a, that is, the flow that reaches the successor node. Therefore,

$$f'_a = \alpha_a f_a, \quad \forall a \in L. \tag{4.3}$$

The number of units of the spoiled produce on link a is the difference between the initial and the final flow, $f_a - f'_a$, where

$$f_a - f'_a = (1 - \alpha_a) f_a, \quad \forall a \in L. \tag{4.4}$$

Associated with the food deterioration is a total discarding cost function, $\hat{z}_a$, which is a function of flow on the link, f_a, that is,

$$\hat{z}_a = \hat{z}_a(f_a), \quad \forall a \in L. \tag{4.5}$$

$\hat{z}_a$ is assumed to be convex and continuously differentiable.

In developed countries, the overall average percentage loss of fruits and vegetables during postproduction supply chain activities is approximately 12% of the initial production. In developing regions the percentage loss is greater. It is imperative to remove the spoiled fresh food products from the supply chain network because the presence of spoiled products may, in many cases, increase the spoilage rate. Here, we focus mainly on the disposal of the decayed food products at the processing, storage, and distribution stages (see also Thompson 2002).

As in Chaps. 2 and 3, x_p represents the (initial) flow of product on path p joining an origin node with a destination node, but the supply chain network is for all firms as in Fig. 4.1. Hence, the origin node i associated with firm i is node i, with a destination node, reflected by the demand market k, being R_k. Since we are dealing with real physical products, the path flows must be nonnegative:

$$x_p \geq 0, \quad \forall p \in \mathscr{P}^i_k; i = 1, \ldots, I; k = 1, \ldots, n_R, \tag{4.6}$$

where $\mathscr{P}^i_k$ is the set of all paths joining the origin node i with destination node R_k, that is, the origin/destination (O/D) pair $w^i_k = (i, R_k)$.

The arc–path multiplier, α_{ap}, which is the product of the multipliers of the links on path p that precede link a on path p is defined analogously in (2.13).

Hence, the relationship between the link flow, f_a, and the path flows can, again, be expressed as

$$f_a = \sum_{i=1}^{I} \sum_{k=1}^{n_R} \sum_{p \in \mathscr{P}^i_k} x_p \alpha_{ap}, \quad \forall a \in L. \tag{4.7}$$

Recall that μ_p denotes the multiplier corresponding to path p, defined as the product of all link multipliers on links comprising that path, that is,

$$\mu_p \equiv \prod_{a \in p} \alpha_a, \quad \forall p \in \mathscr{P}^i_k; i = 1, \ldots, I; k = 1, \ldots, n_R. \tag{4.8}$$

4.2 The Food Supply Chain Network Oligopoly Model

The demand for firm i's product at demand market R_k is a variable and is denoted by d_{ik}. It must be equal to the sum of all the final flows—subject to perishability—on paths connecting O/D pair w_k^i:

$$\sum_{p \in \mathscr{P}_k^i} x_p \mu_p = d_{ik}, \quad i = 1, \ldots, I; k = 1, \ldots, n_R. \tag{4.9}$$

The consumers may differentiate the fresh food products due to food safety and health concerns. We group the demands $d_{ik}; i = 1, \ldots, I; k = 1, \ldots, n_R$ into the $I \times n_R$-dimensional vector d.

4.2.2 Competitive Behavior and Cournot–Nash Equilibrium

We denote the demand price associated with food firm i's product at demand market R_k by ρ_{ik} and assume that

$$\rho_{ik} = \rho_{ik}(d), \quad i = 1, \ldots, I; k = 1, \ldots, n_R. \tag{4.10}$$

Note that the price of food firm i's product at a particular demand market may depend not only on the demands for its product at that and the other demand markets but also on the demands for the other substitutable food products at all the demand markets. These demand price functions are assumed to be continuous, continuously differentiable, and monotone decreasing.

To address the competition among the firms for resources used in the production, processing, storage, and distribution of the fresh produce, we assume that the total operational cost on link a may, in general, depend on the product flows on all the links, that is,

$$\hat{c}_a = \hat{c}_a(f), \quad \forall a \in L, \tag{4.11}$$

where f is the vector of all the link flows. The total cost on each link is assumed to be convex and continuously differentiable.

Let X_i denote the vector of path flows associated with firm i where $X_i \equiv \{\{x_p\} | p \in \mathscr{P}^i\} \in R_+^{n_{\mathscr{P}^i}}$, $\mathscr{P}^i \equiv \cup_{k=1,\ldots,n_R} \mathscr{P}_k^i$, and $n_{\mathscr{P}^i}$ denotes the number of paths from firm i to the demand markets. Thus, X is the vector of all the food firms' strategies, that is, $X \equiv \{\{X_i\} | i = 1, \ldots, I\}$.

The profit function of a food firm is defined as the difference between its revenue and its total costs (operational and discarding). Each firm i seeks to maximize its profit. The statement of the maximization of profits for firm i, in link flows, is

$$\text{Maximize} \quad \sum_{k=1}^{n_R} \rho_{ik}(d) d_{ik} - \sum_{a \in L^i} \left(\hat{c}_a(f) + \hat{z}_a(f_a) \right). \tag{4.12}$$

In view of (4.9), we may redefine the demand price functions (4.10) as $\hat{\rho}_{ik}(x) \equiv \rho_{ik}(d)$, $i = 1,\ldots,I; k = 1,\ldots,n_R$. The statement equivalent to (4.12), but purely in path flows, is

$$\text{Maximize} \sum_{k=1}^{n_R} \left(\hat{\rho}_{ik}(x) \sum_{p \in \mathscr{P}_k^i} x_p \mu_p \right) - \sum_{p \in \mathscr{P}^i} \left(\hat{C}_p(x) + \hat{Z}_p(x) \right), \quad (4.13)$$

where the path total operational costs and path total discarding costs in (4.13) are defined as in (3.19) [see also (3.20)] in Chap. 3, but with the paths corresponding to those in the food supply chain network in Fig. 4.1.

We let $U_i(X)$ denote the profit expression for firm i, as in (4.13), with the I-dimensional vector U being the vector of the profits of all the firms as follows:

$$U = U(X). \quad (4.14)$$

In the Cournot–Nash oligopolistic market framework, each firm selects its product path flows in a noncooperative manner, seeking to maximize its own profit, until an equilibrium is achieved.

Definition 4.1 (Supply Chain Network Cournot–Nash Equilibrium). A path flow pattern $X^* \in K = \prod_{i=1}^I K_i$ constitutes a supply chain network Cournot–Nash equilibrium if for each firm i, $i = 1,\ldots,I$:

$$U_i(X_i^*, \hat{X}_i^*) \geq U_i(X_i, \hat{X}_i^*), \quad \forall X_i \in K_i, \quad (4.15)$$

where $\hat{X}_i^* \equiv (X_1^*,\ldots,X_{i-1}^*, X_{i+1}^*,\ldots,X_I^*)$ and $K_i \equiv \{X_i | X_i \in R_+^{n_{\mathscr{P}^i}}\}$.

In other words, an equilibrium is established if no firm can unilaterally improve upon its profit by changing its product flows in its supply chain network, given the product flow decisions of the other firms.

Next, we provide the variational inequality formulations of the Cournot–Nash equilibrium for the fresh produce supply chain network under oligopolistic competition satisfying Definition 4.1, in terms of both path flows and link flows (see Cournot 1838; Nash 1950, 1951; Gabay and Moulin 1980, and Nagurney 2006).

Theorem 4.1 (Variational Inequality Formulations). *Assume that, for each food firm i, the profit function $U_i(X)$ is concave with respect to the variables in X_i and is continuous and continuously differentiable. Then $X^* \in K$ is a supply chain network Cournot–Nash equilibrium according to Definition 4.1 if and only if it satisfies the variational inequality:*

$$-\sum_{i=1}^{I} \langle \nabla_{X_i} U_i(X^*), X_i - X_i^* \rangle \geq 0, \quad \forall X \in K, \quad (4.16)$$

where $\langle \cdot, \cdot \rangle$ denotes the inner product in the corresponding Euclidean space and $\nabla_{X_i} U_i(X)$ denotes the gradient of $U_i(X)$ with respect to X_i. Variational inequality (4.16), in turn, for our model, is equivalent to the variational inequality: determine the vector of equilibrium path flows $x^* \in K^1$ such that

4.2 The Food Supply Chain Network Oligopoly Model

$$\sum_{i=1}^{I}\sum_{k=1}^{n_R}\sum_{p\in\mathscr{P}_k^i}\left[\frac{\partial \hat{C}_p(x^*)}{\partial x_p}+\frac{\partial \hat{Z}_p(x^*)}{\partial x_p}-\hat{\rho}_{ik}(x^*)\mu_p-\sum_{l=1}^{n_R}\frac{\partial \hat{\rho}_{il}(x^*)}{\partial x_p}\sum_{q\in\mathscr{P}_l^i}\mu_q x_q^*\right]$$
$$\times [x_p - x_p^*] \geq 0, \quad \forall x \in K^1, \tag{4.17}$$

where $K^1 \equiv \{x | x \in R_+^{n_{\mathscr{P}}}\}$, and for each path p, $p \in \mathscr{P}_k^i$; $i = 1,\ldots,I$; $k = 1,\ldots,n_R$,

$$\frac{\partial \hat{C}_p(x)}{\partial x_p} \equiv \frac{\partial (\sum_{q\in\mathscr{P}^i} \hat{C}_q)}{\partial x_p} = \sum_{a\in L^i}\sum_{b\in L^i}\frac{\partial \hat{c}_b(f)}{\partial f_a}\alpha_{ap},$$

$$\frac{\partial \hat{Z}_p(x)}{\partial x_p} \equiv \frac{\partial (\sum_{q\in\mathscr{P}^i} \hat{Z}_q)}{\partial x_p} = \sum_{a\in L^i}\frac{\partial \hat{z}_a(f)}{\partial f_a}\alpha_{ap}, \quad \frac{\partial \hat{\rho}_{il}(x)}{\partial x_p} \equiv \frac{\partial \rho_{il}(d)}{\partial d_{ik}}\mu_p. \tag{4.18}$$

Variational inequality (4.17) can also be reexpressed in terms of link flows as follows: determine the vector of equilibrium link flows and the vector of equilibrium demands $(f^*,d^*) \in K^2$ such that

$$\sum_{i=1}^{I}\sum_{a\in L^i}\left[\sum_{b\in L^i}\frac{\partial \hat{c}_b(f^*)}{\partial f_a}+\frac{\partial \hat{z}_a(f_a^*)}{\partial f_a}\right]\times [f_a - f_a^*]$$
$$+\sum_{i=1}^{I}\sum_{k=1}^{n_R}\left[-\rho_{ik}(d^*)-\sum_{l=1}^{n_R}\frac{\partial \rho_{il}(d^*)}{\partial d_{ik}}d_{il}^*\right]\times [d_{ik} - d_{ik}^*] \geq 0, \quad \forall (f,d) \in K^2, \tag{4.19}$$

where $K^2 \equiv \{(f,d) | \exists x \geq 0$ and (4.7) and (4.9) hold$\}$.

Proof. Variational inequality (4.16) follows directly from Gabay and Moulin (1980). Note that

$$\nabla_{X_i}U_i(X) = \left[\frac{\partial U_i}{\partial x_p}; p \in \mathscr{P}_k^i; k = 1,\ldots,n_R\right]. \tag{4.20}$$

For each path p; $p \in \mathscr{P}_k^i$, we have

$$\frac{\partial U_i}{\partial x_p} = \frac{\partial \left[\sum_{l=1}^{n_R}\left(\hat{\rho}_{il}(x)\sum_{q\in\mathscr{P}_l^i}x_q\mu_q\right)-\sum_{q\in\mathscr{P}^i}\left(\hat{C}_q(x)+\hat{Z}_q(x)\right)\right]}{\partial x_p}$$

$$= \sum_{l=1}^{n_R}\frac{\partial \left[\hat{\rho}_{il}(x)\sum_{q\in\mathscr{P}_l^i}x_q\mu_q\right]}{\partial x_p} - \frac{\partial \left[\sum_{q\in\mathscr{P}^i}\hat{C}_q(x)\right]}{\partial x_p} - \frac{\partial \left[\sum_{q\in\mathscr{P}^i}\hat{Z}_q(x)\right]}{\partial x_p}$$

$$= \hat{\rho}_{ik}(x)\mu_p + \sum_{l=1}^{n_R}\frac{\partial \hat{\rho}_{il}(x)}{\partial x_p}\sum_{q\in\mathscr{P}_l^i}x_q\mu_q - \frac{\partial \left[\sum_{q\in\mathscr{P}^i}\hat{C}_q(x)\right]}{\partial x_p} - \frac{\partial \left[\sum_{q\in\mathscr{P}^i}\hat{Z}_q(x)\right]}{\partial x_p}.$$

$$\tag{4.21}$$

In particular, we have

$$\frac{\partial [\sum_{q\in \mathcal{P}^i} \hat{C}_q(x)]}{\partial x_p} = \frac{\partial [\sum_{b\in L^i} \hat{c}_b(f)]}{\partial x_p} = \sum_{a\in L^i} \frac{\partial [\sum_{b\in L^i} \hat{c}_b(f)]}{\partial f_a} \frac{\partial f_a}{\partial x_p}$$

$$= \sum_{a\in L^i} \sum_{b\in L^i} \frac{\partial \hat{c}_b(f)}{\partial f_a} \alpha_{ap}, \quad (4.22a)$$

$$\frac{\partial [\sum_{q\in \mathcal{P}^i} \hat{Z}_q(x)]}{\partial x_p} = \frac{\partial [\sum_{b\in L^i} \hat{z}_b(f_b)]}{\partial x_p} = \sum_{a\in L^i} \frac{\partial [\sum_{b\in L^i} \hat{z}_b(f_b)]}{\partial f_a} \frac{\partial f_a}{\partial x_p}$$

$$= \sum_{a\in L^i} \frac{\partial \hat{z}_a(f_a)}{\partial f_a} \alpha_{ap}. \quad (4.22b)$$

Therefore, variational inequality (4.17) is established. Also, using Eqs. (4.7) and (4.9), variational inequality (4.19) then follows from (4.17). □

Variational inequalities (4.17) and (4.19) can be put into standard form (2.41). We have already defined X as the vector of path flows (strategies). We now define

$$F(X) \equiv \left[\frac{\partial \hat{C}_p(x)}{\partial x_p} + \frac{\partial \hat{Z}_p(x)}{\partial x_p} - \hat{\rho}_{ik}(x)\mu_p - \sum_{l=1}^{n_R} \frac{\partial \hat{\rho}_{il}(x)}{\partial x_p} \sum_{q\in \mathcal{P}^i_l} \mu_q x_q; \right.$$

$$\left. p \in \mathcal{P}^i_k; i = 1,\ldots,I; k = 1,\ldots,n_R \right], \quad (4.23)$$

and $\mathcal{K} \equiv K^1$ then (4.17) can be reexpressed as (2.41). Similarly, for the variational inequality in terms of link flows, if we define the vectors $X \equiv (f,d)$ and $F(X) \equiv (F_1(X), F_2(X))$, where

$$F_1(X) = \left[\sum_{b\in L^i} \frac{\partial \hat{c}_b(f)}{\partial f_a} + \frac{\partial \hat{z}_a(f_a)}{\partial f_a}; a \in L^i; i = 1,\ldots,I \right],$$

$$F_2(X) = \left[-\rho_{ik}(d) - \sum_{l=1}^{n_R} \frac{\partial \rho_{il}(d)}{\partial d_{ik}} d_{il}; i = 1,\ldots,I; k = 1,\ldots,n_R \right], \quad (4.24)$$

and $\mathcal{K} \equiv K^2$, then (4.19) can be rewritten as (2.41).

Since the feasible sets K^1 and K^2 are not compact, we cannot obtain the existence of a solution simply based on the assumption of the continuity of F. However, the demand d_{ik} for each food firm i's product at every demand market R_k may be assumed to be bounded, since the total demand for these products is finite (although it may be large). Consequently, we have that

$$\mathcal{K}_b \equiv \{x | 0 \leq x \leq b\}, \quad (4.25)$$

where $b > 0$ and $x \leq b$ means that $x_p \leq b$ for all $p \in \mathcal{P}^i_k$; $i = 1,\ldots,I$, and $k = 1,\ldots,n_R$. Then $\mathcal{K}_b$ is a bounded, closed, and convex subset of K^1. Thus, the following variational inequality

4.3 The Algorithm

$$\langle F(X^b), X - X^b \rangle \geq 0, \quad \forall X \in \mathcal{K}_b \tag{4.26}$$

admits at least one solution $X^b \in \mathcal{K}_b$, since $\mathcal{K}_b$ is compact and F is continuous. Therefore, following Kinderlehrer and Stampacchia (1980) (see also Theorem 1.5 in Nagurney 1999), we have the following theorem:

Theorem 4.2 (Existence). *There exists at least one solution to variational inequality (4.17) (also to (4.19)), since there exists a $b > 0$ such that variational inequality (4.26) admits a solution in $\mathcal{K}_b$ with*

$$x^b \leq b. \tag{4.27}$$

In addition, we now provide a uniqueness result.

Theorem 4.3 (Uniqueness). *With Theorem 4.2, variational inequality (4.17) and, hence, variational inequality (4.19) admit at least one solution. Moreover, if the function $F(X)$ of variational inequality (4.19), as defined in (4.24), is strictly monotone on $\mathcal{K} \equiv K^2$, that is,*

$$\langle (F(X^1) - F(X^2)), X^1 - X^2 \rangle > 0, \quad \forall X^1, X^2 \in \mathcal{K}, X^1 \neq X^2, \tag{4.28}$$

then the solution to variational inequality (4.19) is unique, that is, the equilibrium link flow pattern and the equilibrium demand pattern are unique.

It is more reasonable that we would have, given the supply chain network topology, uniqueness of an equilibrium in link flows than in path flows.

Our proposed supply chain network model can also be applied to other fresh food supply chain oligopoly problems under quality competition. In such cases, the food products get delivered to the demand markets with distinct levels of quality degradation. Thus, the arc multiplier for a postproduction link, α_a, captures the corresponding food quality degradation associated with that link, instead of the fraction of unspoiled products. We refer to Labuza (1982) and Man and Jones (1994) for thorough discussions about food quality deterioration.

4.3 The Algorithm

To computationally solve the supply chain network problem, we utilize the Euler method as given in (2.59). Its realization in the context of our competitive food supply chain network problem, governed by the variational inequality formulation (4.17), yields subproblems, at each iteration, that can be computed explicitly in closed form (as was also the case for the blood supply chain network problem in Chap. 2).

4.3.1 Explicit Formulae for the Euler Method Applied to the Fresh Produce Supply Chain Network Oligopoly Variational Inequality (4.17)

The elegance of this procedure for the computation of solutions to the fresh produce supply chain network oligopoly problem can be seen in the following explicit formulae. In particular, we have the following closed form expression for the path flow on each path $p \in \mathscr{P}_k^i; i = 1, \ldots, I; k = 1, \ldots, n_R$, at iteration $\tau + 1$:

$$x_p^{\tau+1} = \max\left\{0, x_p^\tau + a_\tau\left(\hat{\rho}_{ik}(x^\tau)\mu_p + \sum_{l=1}^{n_R} \frac{\partial \hat{\rho}_{il}(x^\tau)}{\partial x_p} \sum_{q \in \mathscr{P}_l^i} \mu_q x_q^\tau - \frac{\partial \hat{C}_p(x^\tau)}{\partial x_p} - \frac{\partial \hat{Z}_p(x^\tau)}{\partial x_p}\right)\right\}. \tag{4.29}$$

It is important to emphasize that the Euler method, which, as noted in Chap. 2, is induced by the general iterative scheme of Dupuis and Nagurney (1993), can also be interpreted as a discrete-time adjustment process. Indeed, the general iterative scheme was devised to not only compute the stationary points of projected dynamical systems (which coincide with the solutions to the associated variational inequality problems) but also provide a means of tracking the associated trajectories over time. Hence, (4.29) may be interpreted as a discrete-time adjustment process with the iteration corresponding to a time period. The firms update their product path flows at each time step by reviewing the product path flows in the previous time period along with the marginal revenue associated with that path. Convergence is achieved to the desired tolerance level when the respective subsequent path flow iterates lie sufficiently near one another, and hence, a stationary point (at which there is essentially no change) has been reached. For a discussion of convergence, see Chap. 2.

In the next section, we solve fresh produce supply chain network oligopoly problems using the above algorithmic scheme.

4.4 A Case Study

This case study focuses on the cantaloupe market in the United States. Most of the cantaloupes consumed in the United States are grown in either California, Mexico, or Central America. We assume that there are two firms, Firm 1 and Firm 2, which may represent, for example, one firm in California and one firm in Central America. Each firm has two production sites, one processor and two distribution centers, and serves two geographically separated demand markets, as depicted in Fig. 4.2. The production sites and the processor of Firm 1 are located in California. The production sites and the processor of Firm 2 are located in Central America, with, typically, lower operational costs. However, all the distribution centers and the demand

4.4 A Case Study

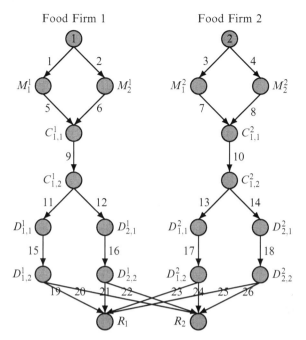

Fig. 4.2 The fresh produce supply chain network topology for the case study

markets are located in the United States. We let D_k^i denote the k-th distribution center of firm i. Hence, as described in Sect. 4.2, $D_{k,1}^i$ denotes the beginning of the storage activity at D_k^i and $D_{k,2}^i$ denotes the end of the storage activity at D_k^i. The first distribution centers of both firms, D_1^1 and D_1^2, are located closer to their respective production sites than their second distribution centers, D_2^1 and D_2^2. The demand market R_1 is located closer to the firms' first distribution centers, D_1^1 and D_1^2, whereas the demand market R_2 is closer to their second distribution centers, D_2^1 and D_2^2.

Typically, cantaloupes can be stored for 12–15 days at 2.2°–5°C (36°–41°F). Their decay may result from postproduction diseases such as Rhizopus, Fusarium, and Geotrichum, depending on the season, the region, and the handling technologies utilized (see Suslow et al. 1997 and Sommer et al. 2002). As discussed in Sect. 4.2, we captured the food deterioration through the arc multipliers. The values of the decay constants and the time durations, although hypothetical, were selected so as to reflect the various supply chain activities. The values of the arc multipliers were calculated using (4.2). For example, we assume that Firm 1 utilizes more effective cleaning and sanitizing techniques for its processing activities which results in relatively higher operational costs but lower decay constants associated with the successive supply chain activities.

We implemented the Euler method (cf. (4.29)) for the solution of variational inequality (4.17), using MATLAB. We set the sequence $\{a_\tau\} = .1\left(1, \frac{1}{2}, \frac{1}{2}, \ldots\right)$. The convergence tolerance was 10^{-6}. In other words, convergence was assumed to have been achieved when the absolute value of the difference between each path flow in two consecutive iterations was less than or equal to this tolerance.

Scenario 1. In Scenario 1, we assumed that consumers at the demand markets are indifferent between the cantaloupes of Firm 1 and Firm 2. Furthermore, consumers at demand market R_2 are willing to pay relatively more than those at demand market R_1. The corresponding demand price functions were

Firm 1: $\rho_{11}(d) = -.0001d_{11} - .0001d_{21} + 4, \rho_{12}(d) = -.0001d_{12} - .0001d_{22} + 6,$
Firm 2: $\rho_{21}(d) = -.0001d_{21} - .0001d_{11} + 4, \rho_{22}(d) = -.0001d_{22} - .0001d_{12} + 6,$

where d_{11}, d_{12}, d_{21}, and d_{22} denote the demands for cantaloupes, per day.

The arc multipliers, the total operational cost functions, and the total discarding cost functions are reported in Table 4.1, as are the decay constants (/day) and the time durations (days) associated with all the links. These cost functions have been constructed based on the data of the average costs available on the web (see, e.g., Meister 2004a,b). Table 4.1 also provides the computed equilibrium product flows on all the links in Fig. 4.2.

The computed equilibrium demands for cantaloupes were

$$d_{11}^* = 7.86, \quad d_{12}^* = 123.62, \quad d_{21}^* = 27.19, \quad d_{22}^* = 139.38.$$

The incurred equilibrium prices were

$$\rho_{11}(d^*) = 4.00, \quad \rho_{12}(d^*) = 5.97, \quad \rho_{21}(d^*) = 4.00, \quad \rho_{22}(d^*) = 5.97.$$

Furthermore, the profits of two firms were

$$U_1(X^*) = 370.46, \quad U_2(X^*) = 454.72.$$

Since consumers do not differentiate between the cantaloupes of the two firms, the prices of these two firms' cantaloupes at each demand market are identical. Due to the difference in consumers' willingness to pay, the price at demand market R_1 is lower than the price at demand market R_2. Consequently, the distribution links 21 and 25, connecting Firm 1 and Firm 2 to demand market R_1, respectively, have zero product flows. In other words, there is no shipment from the distribution centers D_2^1 and D_2^2 to demand market R_1. In addition, the volume of the product on distribution link 22 (or link 26) is higher than that on distribution link 20 (or link 24), which indicates that it is more cost-effective to provide fresh fruits from the nearby distribution centers. As a result of its lower operational costs, Firm 2 dominates both of the demand markets, leading to a substantially higher profit.

Scenario 2. In Scenario 2, we considered that the Centers for Disease Control (CDC) reports a multistate cantaloupe-associated disease outbreak. Due to food safety and health concerns, many of the regular consumers of cantaloupes switch to other fresh fruits. The demand price functions are now given by

4.4 A Case Study

Table 4.1 Decay constants, time durations, arc multipliers, total operational cost and total discarding cost functions, and equilibrium link flow solution for Scenario 1

Link a	λ_a	t_a	α_a	$\hat{c}_a(f)$	$\hat{z}_a(f_a)$	f_a^*
1	–	–	1.00	$.005f_1^2 + .03f_1$	0.00	76.32
2	–	–	1.00	$.006f_2^2 + .02f_2$	0.00	75.73
3	–	–	1.00	$.001f_3^2 + .02f_3$	0.00	103.74
4	–	–	1.00	$.001f_4^2 + .02f_4$	0.00	105.62
5	.150	0.20	.970	$.003f_5^2 + .01f_5$	0.00	76.32
6	.150	0.25	.963	$.002f_6^2 + .02f_6$	0.00	75.73
7	.150	0.30	.956	$.001f_7^2 + .02f_7$	0.00	103.74
8	.150	0.30	.956	$.001f_8^2 + .01f_8$	0.00	105.62
9	.040	0.50	.980	$.002f_9^2 + .05f_9$	$.001f_9^2 + 0.02f_9$	147.01
10	.060	0.50	.970	$.001f_{10}^2 + .02f_{10}$	$.001f_{10}^2 + 0.02f_{10}$	200.14
11	.015	1.50	.978	$.005f_{11}^2 + .01f_{11}$	0.00	65.98
12	.015	3.00	.956	$.01f_{12}^2 + .01f_{12}$	0.00	78.12
13	.025	2.00	.951	$.005f_{13}^2 + .02f_{13}$	0.00	96.47
14	.025	4.00	.905	$.01f_{14}^2 + .01f_{14}$	0.00	97.76
15	.010	3.00	.970	$.004f_{15}^2 + .01f_{15}$	$.001f_{15}^2 + 0.02f_{15}$	64.51
16	.010	3.00	.970	$.004f_{16}^2 + .01f_{16}$	$.001f_{16}^2 + 0.02f_{16}$	74.68
17	.015	3.00	.956	$.004f_{17}^2 + .01f_{17}$	$.001f_{17}^2 + 0.02f_{17}$	91.77
18	.015	3.00	.956	$.004f_{18}^2 + .01f_{18}$	$.001f_{18}^2 + 0.02f_{18}$	88.45
19	.015	1.00	.985	$.005f_{19}^2 + .01f_{19}$	$.001f_{19}^2 + 0.02f_{19}$	7.98
20	.015	3.00	.956	$.015f_{20}^2 + .1f_{20}$	$.001f_{20}^2 + 0.02f_{20}$	54.62
21	.015	3.00	.956	$.015f_{21}^2 + .1f_{21}$	$.001f_{21}^2 + 0.02f_{21}$	0.00
22	.015	1.00	.985	$.005f_{22}^2 + .01f_{22}$	$.001f_{22}^2 + 0.02f_{22}$	72.48
23	.020	1.00	.980	$.005f_{23}^2 + .01f_{23}$	$.001f_{23}^2 + 0.02f_{23}$	27.74
24	.020	3.00	.942	$.015f_{24}^2 + .1f_{24}$	$.001f_{24}^2 + 0.02f_{24}$	59.99
25	.020	3.00	.942	$.015f_{25}^2 + .1f_{25}$	$.001f_{25}^2 + 0.02f_{25}$	0.00
26	.020	1.00	.980	$.005f_{26}^2 + .01f_{26}$	$.001f_{26}^2 + 0.02f_{26}$	84.56

Firm 1: $\rho_{11}(d) = -.001d_{11} - .001d_{21} + .5$, $\quad \rho_{12}(d) = -.001d_{12} - .001d_{22} + .5$,

Firm 2: $\rho_{21}(d) = -.001d_{21} - .001d_{11} + .5$, $\quad \rho_{22}(d) = -.001d_{22} - .001d_{12} + .5$.

The longer shipment times associated with links 13 and 14 in Table 4.2 represent the need for inspections of imported food by the US government. Therefore, the values of arc multipliers associated with links 13 and 14 in Table 4.2 are lower than those in Table 4.1, which implies more cantaloupes will spoil during these links. The other arc multipliers and the total operational and the total discarding cost functions are the same as in Scenario 1, as shown in Table 4.2. The new computed equilibrium link flows are also reported in Table 4.2.

The computed equilibrium demands for cantaloupes were

$$d_{11}^* = 4.51, \quad d_{12}^* = 3.24, \quad d_{21}^* = 5.96, \quad d_{22}^* = 4.21.$$

The incurred equilibrium prices were:

$$\rho_{11}(d^*) = 0.49, \quad \rho_{12}(d^*) = 0.49, \quad \rho_{21}(d^*) = 0.49, \quad \rho_{22}(d^*) = 0.49.$$

Table 4.2 Decay constants, time durations, arc multipliers, total operational cost and total discarding cost functions, and equilibrium link flow solution for Scenario 2

Link a	λ_a	t_a	α_a	$\hat{c}_a(f)$	$\hat{z}_a(f_a)$	f_a^*
1	–	–	1.00	$.005f_1^2 + .03f_1$	0.00	4.43
2	–	–	1.00	$.006f_2^2 + .02f_2$	0.00	4.40
3	–	–	1.00	$.001f_3^2 + .02f_3$	0.00	5.94
4	–	–	1.00	$.001f_4^2 + .02f_4$	0.00	6.94
5	.150	0.20	.970	$.003f_5^2 + .01f_5$	0.00	4.43
6	.150	0.25	.963	$.002f_6^2 + .02f_6$	0.00	4.40
7	.150	0.30	.956	$.001f_7^2 + .02f_7$	0.00	5.94
8	.150	0.30	.956	$.001f_8^2 + .01f_8$	0.00	6.94
9	.040	0.50	.980	$.002f_9^2 + .05f_9$	$.001f_9^2 + 0.02f_9$	8.53
10	.060	0.50	.970	$.001f_{10}^2 + .02f_{10}$	$.001f_{10}^2 + 0.02f_{10}$	12.31
11	.015	1.50	.978	$.005f_{11}^2 + .01f_{11}$	0.00	4.82
12	.015	3.00	.956	$.01f_{12}^2 + .01f_{12}$	0.00	3.54
13	.025	3.00	.928	$.005f_{13}^2 + .02f_{13}$	0.00	6.86
14	.025	5.00	.882	$.01f_{14}^2 + .01f_{14}$	0.00	5.09
15	.010	3.00	.970	$.004f_{15}^2 + .01f_{15}$	$.001f_{15}^2 + 0.02f_{15}$	4.72
16	.010	3.00	.970	$.004f_{16}^2 + .01f_{16}$	$.001f_{16}^2 + 0.02f_{16}$	3.38
17	.015	3.00	.956	$.004f_{17}^2 + .01f_{17}$	$.001f_{17}^2 + 0.02f_{17}$	6.36
18	.015	3.00	.956	$.004f_{18}^2 + .01f_{18}$	$.001f_{18}^2 + 0.02f_{18}$	4.49
19	.015	1.00	.985	$.005f_{19}^2 + .01f_{19}$	$.001f_{19}^2 + 0.02f_{19}$	4.58
20	.015	3.00	.956	$.015f_{20}^2 + .1f_{20}$	$.001f_{20}^2 + 0.02f_{20}$	0.00
21	.015	3.00	.956	$.015f_{21}^2 + .1f_{21}$	$.001f_{21}^2 + 0.02f_{21}$	0.00
22	.015	1.00	.985	$.005f_{22}^2 + .01f_{22}$	$.001f_{22}^2 + 0.02f_{22}$	3.28
23	.020	1.00	.980	$.005f_{23}^2 + .01f_{23}$	$.001f_{23}^2 + 0.02f_{23}$	6.08
24	.020	3.00	.942	$.015f_{24}^2 + .1f_{24}$	$.001f_{24}^2 + 0.02f_{24}$	0.00
25	.020	3.00	.942	$.015f_{25}^2 + .1f_{25}$	$.001f_{25}^2 + 0.02f_{25}$	0.00
26	.020	1.00	.980	$.005f_{26}^2 + .01f_{26}$	$.001f_{26}^2 + 0.02f_{26}$	4.29

Furthermore, the profits of two firms were:

$$U_1(X^*) = 1.16, \quad U_2(X^*) = 1.63.$$

The demand for cantaloupes is decreased by the associated outbreak, significantly decreasing the demand prices at demand markets. Both firms experience dramatic declines in their profits. In addition, additional distribution links 20, 21, 24, and 25 have zero product flows (as compared to Scenario 1), since the extremely low demand price cannot cover the costs associated with long-distance distribution.

Scenario 3. Given the severe shrinkage in the demand, Firm 1 has realized the importance of regaining consumers' confidence in its own product after the outbreak. Thus, Firm 1 has the label on its cantaloupes redesigned to incorporate a guarantee of food safety. This causes additional expenditures associated with its processing activities. The arc multipliers and the total operational and the total discarding cost functions are the same as in Scenario 2, except for the total operational cost function associated with the processing link 9. Please refer to Table 4.3.

4.4 A Case Study

Table 4.3 Decay constants, time durations, multipliers, total operational cost and total discarding cost functions, and equilibrium link flow solution for Scenario 3

Link a	λ_a	t_a	α_a	$\hat{c}_a(f)$	$\hat{z}_a(f_a)$	f_a^*
1	–	–	1.00	$.005 f_1^2 + .03 f_1$	0.00	36.92
2	–	–	1.00	$.006 f_2^2 + .02 f_2$	0.00	36.64
3	–	–	1.00	$.001 f_3^2 + .02 f_3$	0.00	5.43
4	–	–	1.00	$.001 f_4^2 + .02 f_4$	0.00	6.44
5	.150	0.20	.970	$.003 f_5^2 + .01 f_5$	0.00	36.92
6	.150	0.25	.963	$.002 f_6^2 + .02 f_6$	0.00	36.64
7	.150	0.30	.956	$.001 f_7^2 + .02 f_7$	0.00	5.43
8	.150	0.30	.956	$.001 f_8^2 + .01 f_8$	0.00	6.44
9	.040	0.50	.980	$.003 f_9^2 + .06 f_9$	$.001 f_9^2 + 0.02 f_9$	71.11
10	.060	0.50	.970	$.001 f_{10}^2 + .02 f_{10}$	$.001 f_{10}^2 + 0.02 f_{10}$	11.35
11	.015	1.50	.978	$.005 f_{11}^2 + .01 f_{11}$	0.00	36.33
12	.015	3.00	.956	$.01 f_{12}^2 + .01 f_{12}$	0.00	33.38
13	.025	3.00	.928	$.005 f_{13}^2 + .02 f_{13}$	0.00	6.68
14	.025	5.00	.882	$.01 f_{14}^2 + .01 f_{14}$	0.00	4.33
15	.010	3.00	.970	$.004 f_{15}^2 + .01 f_{15}$	$.001 f_{15}^2 + 0.02 f_{15}$	35.52
16	.010	3.00	.970	$.004 f_{16}^2 + .01 f_{16}$	$.001 f_{16}^2 + 0.02 f_{16}$	31.91
17	.015	3.00	.956	$.004 f_{17}^2 + .01 f_{17}$	$.001 f_{17}^2 + 0.02 f_{17}$	6.20
18	.015	3.00	.956	$.004 f_{18}^2 + .01 f_{18}$	$.001 f_{18}^2 + 0.02 f_{18}$	3.82
19	.015	1.00	.985	$.005 f_{19}^2 + .01 f_{19}$	$.001 f_{19}^2 + 0.02 f_{19}$	17.78
20	.015	3.00	.956	$.015 f_{20}^2 + .1 f_{20}$	$.001 f_{20}^2 + 0.02 f_{20}$	16.69
21	.015	3.00	.956	$.015 f_{21}^2 + .1 f_{21}$	$.001 f_{21}^2 + 0.02 f_{21}$	0.00
22	.015	1.00	.985	$.005 f_{22}^2 + .01 f_{22}$	$.001 f_{22}^2 + 0.02 f_{22}$	30.96
23	.020	1.00	.980	$.005 f_{23}^2 + .01 f_{23}$	$.001 f_{23}^2 + 0.02 f_{23}$	5.93
24	.020	3.00	.942	$.015 f_{24}^2 + .1 f_{24}$	$.001 f_{24}^2 + 0.02 f_{24}$	0.00
25	.020	3.00	.942	$.015 f_{25}^2 + .1 f_{25}$	$.001 f_{25}^2 + 0.02 f_{25}$	0.00
26	.020	1.00	.980	$.005 f_{26}^2 + .01 f_{26}$	$.001 f_{26}^2 + 0.02 f_{26}$	3.65

The demand price functions corresponding to the two demand markets for cantaloupes from these two firms are now given by

Firm 1: $\rho_{11}(d) = -.001 d_{11} - .0005 d_{21} + 2.5$, $\rho_{12}(d) = -.0003 d_{12} - .0002 d_{22} + 3$,

Firm 2: $\rho_{21}(d) = -.001 d_{21} - .001 d_{11} + .5$, $\rho_{22}(d) = -.001 d_{22} - .001 d_{12} + .5$.

The computed values of the equilibrium link flows are given in Table 4.3.

Note that links 21, 24, and 25 have zero flow. Hence, as we did for the medical nuclear supply chain network case study in Chap. 3, we depict the *final* network topology for this scenario with only the links with positive equilibrium flows displayed in Fig. 4.3. Note that the analogous *final* topology for Scenario 2 would be that given in Fig. 4.3 but with link 20 removed. As for the *final* topology for Scenario 1, it would be as in Fig. 4.2 but with only links 21 and 25 removed, since those have zero flow at the equilibrium solution.

The computed equilibrium demands for cantaloupes were

$$d_{11}^* = 17.52, \quad d_{12}^* = 46.46, \quad d_{21}^* = 5.81, \quad d_{22}^* = 3.58.$$

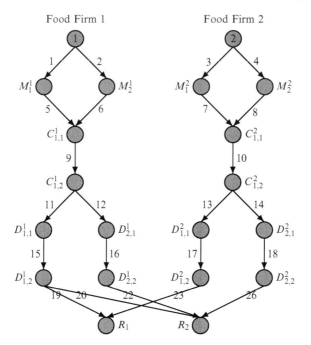

Fig. 4.3 The *final* fresh produce supply chain network topology for Scenario 3 with only links with positive equilibrium flows displayed

The incurred equilibrium prices at the demand markets were

$$\rho_{11}(d^*) = 2.48, \quad \rho_{12}(d^*) = 2.99, \quad \rho_{21}(d^*) = 0.48, \quad \rho_{22}(d^*) = 0.45.$$

Furthermore, the profits of two firms were

$$U_1(X^*) = 84.20, \quad U_2(X^*) = 1.38.$$

Consumers differentiate cantaloupes due to food safety and health concerns in Scenario 3. With the newly designed label, Firm 1 has managed to increase the consumption of its cantaloupes at both demand markets, whereas the demands for Firm 2's cantaloupes are even lower than those in Scenario 2. Because of the cantaloupe-associated foodborne disease outbreak, the firms do not return to the same profit level as in Scenario 1. A comparison of the results in Scenario 2 and Scenario 3 suggests that practicing product differentiation may be an effective strategy for a food firm to maintain its profit. The demand for Firm 1's product at demand market R_1 in Scenario 3 is even higher than that in Scenario 1, which is likely due to the decrease in the price coupled with the introduced guarantee of food safety.

It is also interesting to note that, due to food deterioration, Firm 1 produced an amount $f_1^* + f_2^* = 73.56$ for the total demand of $d_{11}^* + d_{12}^* = 63.98$, with roughly 13% of the cantaloupes spoiled and discarded during the postproduction supply chain activities. Firm 2 produced an amount of $f_3^* + f_4^* = 11.87$ for the total demand of $d_{21}^* + d_{22}^* = 9.39$.

Table 4.4 Computed equilibrium path flow pattern for Scenario 3

	Path definition	Path flow
O/D pair $w_1^1 = (1, R_1)$	$p_1 = (1, 5, 9, 11, 15, 19)$	$x_{p_1}^* = 9.93$
	$p_2 = (1, 5, 9, 12, 16, 21)$	$x_{p_2}^* = 0.00$
	$p_3 = (2, 6, 9, 11, 15, 19)$	$x_{p_3}^* = 9.85$
	$p_4 = (2, 6, 9, 12, 16, 21)$	$x_{p_4}^* = 0.00$
O/D pair $w_2^1 = (1, R_2)$	$p_1 = (1, 5, 9, 11, 15, 20)$	$x_{p_1}^* = 9.32$
	$p_2 = (1, 5, 9, 12, 16, 22)$	$x_{p_2}^* = 17.68$
	$p_3 = (2, 6, 9, 11, 15, 20)$	$x_{p_3}^* = 9.24$
	$p_4 = (2, 6, 9, 12, 16, 22)$	$x_{p_4}^* = 17.54$
O/D pair $w_1^2 = (2, R_1)$	$p_1 = (3, 7, 10, 13, 17, 23)$	$x_{p_1}^* = 3.35$
	$p_2 = (3, 7, 10, 14, 18, 25)$	$x_{p_2}^* = 0.00$
	$p_3 = (4, 8, 10, 13, 17, 23)$	$x_{p_3}^* = 3.85$
	$p_4 = (4, 8, 10, 14, 18, 25)$	$x_{p_4}^* = 0.00$
O/D pair $w_2^2 = (2, R_2)$	$p_1 = (3, 7, 10, 13, 17, 24)$	$x_{p_1}^* = 0.00$
	$p_2 = (3, 7, 10, 14, 18, 26)$	$x_{p_2}^* = 2.08$
	$p_3 = (4, 8, 10, 13, 17, 24)$	$x_{p_3}^* = 0.00$
	$p_4 = (4, 8, 10, 14, 18, 26)$	$x_{p_4}^* = 2.58$

As we did in Chap. 3, we now also present the computed equilibrium path flows. In Table 4.4, we report both nonzero (positive) and zero path flows since there are fewer paths than in the Scenario 3 example in Chap. 3. Links are numbered as in Fig. 4.2. Only two of the available four paths connecting each of the O/D pairs, w_1^1, w_1^2, and w_2^2, are used at the equilibrium solution. Only O/D pair w_2^1 has all its paths used.

Remark. We emphasize that, with appropriate modifications, the model can handle the degradation in quality, which is more relevant to meat, dairy, and bakery products, and even quality competition. For example, the path multipliers μ_p; $p \in \mathscr{P}$ can be used as quality parameters in the demand price functions. The conservation of flow equations would need to be then modified, since the arc and path multipliers would not measure quantity losses. For a dynamic network oligopoly model with quality competition and product differentiation, see Nagurney and Li (2012).

4.5 Summary and Conclusions

This chapter focused on food spoilage/deterioration between production and consumption locations, which poses unique challenges for food supply chain management. In particular, we presented a food supply chain network model under oligopolistic competition and perishability, with a focus on fresh produce, such as vegetables and fruits. Each food firm is involved in such supply chain activities as the production, processing, storage, distribution, and even the disposal of the food products and seeks to determine its optimal product flows, in order to maximize its own profit.

We captured the exponential time decay of food in the number of units through the introduction of arc multipliers, which depend on the time duration and the environmental conditions associated with each postproduction supply chain activity. We also incorporated the discarding costs associated with the disposal of the spoiled food products at the processing, storage, and distribution stages. Moreover, the competitive model allows consumers to differentiate food products at the demand markets due to product freshness and food safety concerns. In addition, the flexibility of the supply chain network topology enables decision-makers to evaluate alternative technologies involved in various supply chain activities.

We derived the variational inequality formulations of the food supply chain network Cournot–Nash equilibrium and presented some qualitative properties of the equilibrium pattern. We also provided the explicit formulae for each step of the iterative scheme, the Euler method, which was also the algorithm used for the blood supply chain network model in Chap. 2. We illustrated the model and the algorithm through a case study of the cantaloupe market in the United States. The results of the case study suggest that product differentiation may be an effective strategy for financial resiliency, especially in times of foodborne disease outbreaks.

4.6 Sources and Notes

This chapter is based on the paper by Yu and Nagurney (2013). Here, we integrate food supply chains into the broader context of perishable product supply chain networks, the focus of this book, report additional output data, and provide an expanded discussion. As noted by Van der Vorst (2006), it is imperative to analyze food supply chains within the full complexity of their network structure, as we have done here. Monteiro (2007) claimed that the theory of network economics (cf. Nagurney 1999) provides a powerful framework in which the supply chain can be graphically represented and analyzed, which we also use as a foundation. He utilized the theory to study the traceability. Blackburn and Scudder (2009) suggested a cost minimization model for specific perishable product supply chain design, capturing the declining value of the product over time. The authors noted that product value deteriorates significantly over time at rates that highly depend on temperature and humidity. Rong (2011) presented a mixed integer linear programming model for food production and distribution planning (see also Kopanos 2012) with a focus on quality.

Several other contributions that integrated and synthesized two or more processes associated with food supply chains are worth mentioning. Zhang et al. (2003) studied a physical distribution system in order to minimize the total cost for storage and shipment with the product quality requirement fulfilled. Widodo et al. (2006) developed models dealing with flowering-harvesting and harvesting-delivering problems of agricultural products by introducing a plant-maturing curve and a loss function to address, respectively, the growing process and the decaying process of the products. Ahumada and Villalobos (2011) discussed the packing and distribution problem of fresh produce, with the inclusion of perishability.

Numerous challenges have underlined the need for the efficient management of food supply chains. The review by Lowe and Preckel (2004) focused on farm planning. Lütke Entrup (2005) discussed how to integrate shelf life into production planning in three sample food industries (yogurt, sausages, and poultry). Akkerman et al. (2010) outlined quantitative operations management applications in food distribution management. The survey by Lucas and Chhajed (2004) presented applications related to location problems in agriculture and recognized the challenges of strategic production-distribution planning problems in this industry. Due to the added complexity of perishability, there are fewer articles related to perishable food products than those related to nonperishable ones and even fewer models developed for fully integrated supply chain systems, the topic of this chapter.

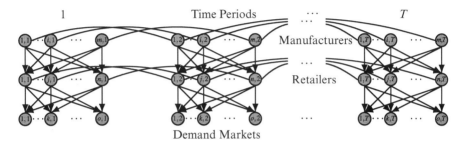

Fig. 4.4 The network structure of the multiperiod oligopolistic supply chain

As noted in Chaps. 1 and 2, our approach to product perishability utilizes generalized network concepts for nonlinear networks, with ideas based on Nagurney and Aronson (1989), who presented a multiperiod spatial price network equilibrium model. Spatial price equilibrium problems reflect perfect competition. In this book, we focus on a finite number of decision-makers, who have control over their supply chains. Liu and Nagurney (2012) proposed (cf. Fig. 4.4) a multiperiod oligopolistic network equilibrium model where perishability of the product over a fixed lifetime was captured through appropriate path definitions, reflecting that a product would not be consumed after a certain finite time period. The authors allowed for competition among the manufacturers and among the retailers, over the finite time horizon. Such supply chain network equilibrium models, also formulated as variational inequality problems, were introduced by Nagurney et al. (2002).

References

Ahumada O, Villalobos JR (2009) Application of planning models in the agri-food supply chain: a review. Eur J Oper Res 196(1):1–20

Ahumada O, Villalobos JR (2011) A tactical model for planning the production and distribution of fresh produce. Ann Oper Res 190:339–358

Aiello G, La Scalia G, Micale R (2012) Simulation analysis of cold chain performance based on time-temperature data. Prod Plann Contr 23(6):468–476

Akkerman R, Farahani P, Grunow M (2010) Quality, safety and sustainability in food distribution: A review of quantitative operations management approaches and challenges. OR Spectrum 32(4):863–904

Bhuyan S (2005) Does vertical integration effect market power? Evidence from U.S. food manufacturing industries. J Agr Appl Econ 37(1):263–276

Blackburn J, Scudder G (2009) Supply chain strategies for perishable products: The case of fresh produce. Prod Oper Manag 18(2):129–137

Cournot AA (1838) Researches into the mathematical principles of the theory of wealth, English translation. MacMillan, London

Dupuis P, Nagurney A (1993) Dynamical systems and variational inequalities. Ann Oper Res 44:9–42

Fu B, Labuza TP (1993) Shelf-life prediction: theory and application. Food Contr 4(3):125–133

Gabay D, Moulin H (1980) On the uniqueness and stability of Nash equilibria in noncooperative games. In: Bensoussan A, Kleindorfer P, Tapiero CS (eds) Applied stochastic control of econometrics and management science. North-Holland, Amsterdam, pp 271–294

Gustavsson J, Cederberg C, Sonesson U, Van Otterdijk R, Meybeck A (2011) Global food losses and food waste. The Food and Agriculture Organization of the United Nations, Rome, Italy

Kinderlehrer D, Stampacchia G (1980) An introduction to variational inequalities and their applications. Academic, New York

Kopanos GM, Puigjaner L, Georgiadis MC (2012) Simultaneous production and logistics operations planning in semicontinuous food industries. Omega 40(5): 634–650

Labuza TP (1982) Shelf-life dating of foods. Food & Nutrition Press, Westport, CT

Liu Z, Nagurney A (2012) Multiperiod competitive supply chain networks with inventorying and a transportation network equilibrium reformulation. Optim Eng 13(3):471–503

Lowe TJ, Preckel PV (2004) Decision technologies for agribusiness problems: A brief review of selected literature and a call for research. Manuf Serv Oper Manag 6(3):201–208

Lucas MT, Chhajed D (2004) Applications of location analysis in agriculture: A survey. J Oper Res Soc 55(6):561–578

Lütke Entrup M (2005) Advanced planning in fresh food industries: integrating shelf life into production planning. Physica-Verlag/Springer, Heidelberg

Man CMD, Jones AA (1994) Shelf life evaluation of foods. Blackie Academic & Professional, Glasgow

Meister HS (2004a) Sample cost to establish and produce cantaloupes (slant-bed, spring planted). U.C. Cooperative Extension – Imperial County Vegetable Crops Guidelines, August 2004

Meister HS (2004b) Sample cost to establish and produce cantaloupes (mid-bed trenched). U.C. Cooperative Extension – Imperial County Vegetable Crops Guidelines, August 2004

Monteiro DMS (2007) Theoretical and empirical analysis of the economics of traceability adoption in food supply chains. PhD Thesis, University of Massachusetts, Amherst, Massachusetts

Nagurney A (1999) Network economics: a variational inequality approach, 2nd and revised edition. Kluwer Academic, Dordrecht

Nagurney A (2006) Supply chain network economics: dynamics of prices, flows and profits. Edward Elgar Publishing, Cheltenham

Nagurney A, Aronson J (1989) A general dynamic spatial price network equilibrium model with gains and losses. Networks 19(7):751–769

Nagurney A, Dong J, Zhang D (2002) A supply chain network equilibrium model. Transport Res E 38(5):281–303

Nagurney A, Li D (2012) A dynamic network oligopoly model with transportation costs, product differentiation, and quality competition. Isenberg School of Management, University of Massachusetts, Amherst, MA

Nahmias S (1982) Perishable inventory theory: a review. Oper Res 30(4):680–708

Nash JF (1950) Equilibrium points in n-person games. Proc Natl Acad Sci USA 36:48–49

Nash JF (1951) Noncooperative games. Ann Math 54:286–298

Nga MTT (2010) Enhancing quality management of fresh fish supply chains through improved logistics and ensured traceability. PhD Thesis, University of Iceland, Reykjavik, Iceland

Rong A, Akkerman R, Grunow M (2011) An optimization approach for managing fresh food quality throughout the supply chain. Int J Prod Econ 131(1):421–429

Sommer NF, Fortlage RJ, Edwards DC (2002) Postharvest diseases of selected commodities. In: Kader AA (ed) Postharvest technology of horticultural crops, 3rd edn. University of California Agriculture & Natural Resources, Publication 3311, Oakland, California, pp 197–249

Suslow TV, Cantwell M, Mitchell J (1997) Cantaloupe: recommendations for maintaining postharvest quality. Department of Vegetable Crops, University of California, Davis, California

Thompson JF (2002) Waste management and cull utilization. In: Kader AA (ed) Postharvest technology of horticultural crops, 3rd edn. University of California Agriculture & Natural Resources, Publication 3311, Oakland, California, pp 81–84

Tijskens LMM, Polderdijk JJ (1996) A generic model for keeping quality of vegetable produce during storage and distribution. Agr Syst 51(4):431–452

United States Department of Agriculture (USDA) (2011) Fruits and vegetables (farm weight): Per capita availability, 1970–2009. Available online at: http://www.ers.usda.gov/datafiles/Food_Availability_Per_Capita_Data_System/Food_Availability/fruitveg.xls

Van der Vorst JGAJ (2006) Performance measurement in agri-food supply-chain networks: An overview. In: Ondersteijn CJM, Wijnands JHM, Huirne RBM, Van

Kooten O (eds) Quantifying the agri-food supply chain. Springer, Dordrecht, pp 15–26

Van Zyl GJJ (1964) Inventory control for perishable commodities. PhD Thesis, University of North Carolina, Chapel Hill, North Carolina

Widodo KH, Nagasawa H, Morizawa K, Ota M (2006) A periodical flowering-harvesting model for delivering agricultural fresh products. Eur J Oper Res 170(1):24–43

Yu M, Nagurney A (2013) Competitive food supply chain networks with application to fresh produce. Eur J Oper Res 224(2):273–282

Zhang G, Habenicht W, SpießWEL (2003) Improving the structure of deep frozen and chilled food chain with tabu search procedure. J Food Eng 60(1):67–79

Chapter 5
Pharmaceutical Supply Chains

Abstract In this chapter, we present a generalized network oligopoly model for supply chains of pharmaceutical products using variational inequality theory. The model captures the Cournot competition among the manufacturers who seek to determine their profit-maximizing product flows, with the consumers differentiating among the products of the firms, whether branded or generic, and the firms taking into consideration the discarding costs. A case study demonstrates that a brand pharmaceutical product may lose its dominant market share as a consequence of patent expiration and because of generic competition.

5.1 Motivation and Overview

In this chapter, we develop a generalized network oligopoly model for pharmaceutical supply chain competition. It takes into account, as the food supply chain network model of Chap. 4 did, product perishability, brand differentiation, and discarding costs. Our generalized network-based framework captures competition among the firms during the various supply chain activities of manufacturing, storage, and distribution. The firms are assumed to not only maximize their own profits but also to minimize the waste discarding costs. Although the underlying behavior of the firms is similar to that of the profit-maximizing firms in Chap. 4, the supply chain network topology is different.

Pharmaceutical manufacturing is an immense global industry. Global spending for medicines is expected to exceed $1 trillion by 2015. However, the current five-year compound annual growth rate of 3–6% represents a significant slowdown from the 6.2% annual growth seen in the previous five years (Zacks Equity Research 2012).

Pharmaceutical supply chains have begun to be coupled with sophisticated technologies in order to improve both the quantity and the quality of their associated products. Despite all the advances in manufacturing, storage, and distribution methods, certain pharmaceutical drug companies are far from effectively satisfying market demands on a consistent basis. It has been argued that pharmaceutical drug supply chains are in urgent need of efficient optimization techniques to reduce costs and to increase productivity and responsiveness (Shah 2004 and Papageorgiou 2009).

Product perishability is another critical issue in pharmaceutical/drug supply chains. In a 2003 survey, the estimated incurred cost due to the expiration of branded pharmaceutical products in supermarkets and drug-stores was over 500 million dollars (Karaesmen et al. 2011). In 2007, in a warehouse belonging to the Health Department of Chicago, over one million dollars in drugs, vaccines, and other medical supplies were found spoiled, stolen, or unaccounted for (Mihalopoulos 2009). In 2009, CVS pharmacies in California, as a result of a settlement of a lawsuit filed against the company, were required to offer promotional coupons to customers who had identified expired products including expired baby formula and children's medicines in more than 42% of their stores (WPRI 2009 and Business Wire 2009). A law was passed in the USA in 1979 requiring drug manufacturers to stamp an expiration date on their products. This is the last date that the manufacturer can still guarantee the full potency and safety of the drug, assuming that proper handling and storage procedures have been followed.

Ironically, whereas some drugs may be unsold and unused and/or past their expiration dates, the number of drugs that were reported in short supply in the USA in the first half of 2011 rose to 211—close to an all-time record—as compared to only 58 in short supply in 2004 (Emanuel 2011). According to the Food and Drug Administration (FDA), hospitals have reported shortages of drugs used in a wide range of applications, including cancer treatment to surgery, anesthesia, and intravenous feedings. The consequences of such shortages include the postponement of surgeries and treatments or in the use of less effective or costlier substitutes. According to the American Hospital Association, all USA hospitals have experienced drug shortages, and 82% have reported delayed care for their patients as a consequence (Szabo 2011).

While the real causes of such shortages are complex, most cases appear to be related to manufacturers' decisions to cease production due to financial considerations. For example, among curative cancer drugs, only the older generic, yet less expensive, ones have experienced shortages. As noted by Shah (2004), pharmaceutical companies secure significant financial returns in the early lifetime of a successful drug before competition occurs. This competition-free timespan has been observed to be shortening, from 5 years to only 1–2 years. Hence, the low profit margins associated with such drugs may be forcing pharmaceutical companies to make a difficult decision: whether to lose money by continuing to produce a lifesaving product or to switch to a more profitable drug. The FDA cannot force companies to continue to produce low-profit medicines even if millions of lives rely on them. On the other hand, where competition has been lacking, shortages of some other lifesaving drugs have resulted in huge spikes in prices, ranging from a 100% to a 4,500% increase (Schneider 2011). Adding to the economic pressures, pharmaceutical companies are expected to suffer a significant decrease in their revenues as a result of expiring patent protection for ten of the best-selling drugs by the end of 2012 (De la Garza 2011). In addition to increasing competition from generic drugs, lower reimbursements by government health programs have worsened the situation.

5.2 The Pharmaceutical Supply Chain Network Oligopoly Model

Beyond the financial pressures and challenges, the safety of imported and outsourced products is another major issue for pharmaceutical companies. The emergence of counterfeit products has resulted in major reforms in the relationships among various tiers in pharmaceutical supply chains (Dunehew 2005). Marucheck et al. (2011) noted that while, in the past, product recalls were mainly related to local errors in design, manufacturing, or labeling, today, a single product safety issue may result in huge global consequences. Currently, more than 80% of the ingredients of drugs sold in the USA are made overseas, mostly in remote facilities located in China and India that are rarely, if ever, visited by government inspectors. Supply chains of generic drugs, which account for 75% of the prescription medicines sold in the USA, are more susceptible to falsification. The supply chains of some of the over-the-counter products, such as vitamins or aspirin, are also vulnerable to adulteration (Harris 2011). Similarly, the amount of counterfeit drugs in European pharmaceutical supply chains has considerably increased (Muller 2009). Another pressure faced by pharmaceutical firms is the environmental impact of their waste, which includes the disposal of expired and damaged drugs.

This chapter is organized as follows. In Sect. 5.2, we develop the pharmaceutical supply chain generalized network oligopoly model with perishability and brand differentiation and derive variational inequality formulations, and we discuss special cases of the model. Qualitative properties are also highlighted. In Sect. 5.3, we present the computational algorithm, which we then apply in a case study in Sect. 5.4. In Sect. 5.5, we discuss how the model, over an expanded topology, can also handle outsourcing options. In Sect. 5.6, we discuss the use of an expanded network topology for mergers and acquisitions. We summarize our results in Sect. 5.7. Sources and Notes follow in Sect. 5.8.

5.2 The Pharmaceutical Supply Chain Network Oligopoly Model

We consider I pharmaceutical firms, with a typical firm denoted by i. The firms compete noncooperatively, in an oligopolistic manner, and the consumers can differentiate among the products through their individual product brands. The supply chain network activities include manufacturing, shipment, storage, and, ultimately, the distribution of the brand name drugs to the demand markets.

Consider the supply chain network topology presented in Fig. 5.1. Each pharmaceutical firm i utilizes n_M^i manufacturing plants and n_D^i distribution/storage facilities. The goal is to serve n_R demand markets consisting of pharmacies, retail stores, hospitals, and other medical centers.

L^i, as in Chap. 4, denotes the set of directed links corresponding to the sequence of activities associated with firm i but now using the topology in Fig. 5.1. $\mathscr{G} = [N,L]$ denotes the graph composed of the set of nodes N and the set of links L, where L contains all sets of L^is: $L \equiv \cup_{i=1,...,I} L^i$.

In Fig. 5.1, the first set of links connecting the top two tiers of nodes corresponds to the production of the drugs at each of the manufacturing units of firm i. Such

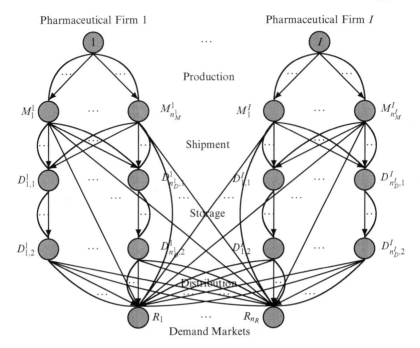

Fig. 5.1 The pharmaceutical supply chain network topology

facilities are denoted by $M_1^i, \ldots, M_{n_M^i}^i$. Note that we allow for multiple possible links connecting each top tier node i with its manufacturing facilities $M_1^i, \ldots, M_{n_M^i}^i$ to represent different possible manufacturing technologies that may be associated with a given facility. We emphasize that the manufacturing facilities may be located not only in different regions of the same country but also in different countries.

The next set of nodes represents the distribution centers, and thus, the links connecting the manufacturing nodes to the distribution centers are shipment-type links. Such distribution nodes associated with firm i are denoted by $D_{1,1}^i, \ldots, D_{n_D^i,1}^i$ and represent the distribution centers that the drugs are shipped to and stored at, before being delivered to the demand markets. There are alternative shipment links to denote different possible modes of transportation. For example, pharmaceuticals that are perishable may be shipped by air, at reduced shipping time, but at a higher cost.

The next set of links connecting nodes $D_{1,1}^i, \ldots, D_{n_D^i,1}^i$ to $D_{1,2}^i, \ldots, D_{n_D^i,2}^i$ represents storage. Since different drugs may require different storage conditions before being ultimately shipped to the demand markets, these alternatives are represented through multiple links at this tier.

The last set of links connecting the two bottom tiers of the supply chain network corresponds to distribution links over which the products are shipped from the distribution/storage facilities to the demand markets. Here multiple modes of shipment/transportation are also allowed.

5.2 The Pharmaceutical Supply Chain Network Oligopoly Model

In addition, in the supply chain network topology in Fig. 5.1, there are direct links connecting manufacturing units with various demand markets in order to capture the possibility of direct shipments from manufacturers.

As in the models in Chaps. 2 through 4, we take into account the perishability of the pharmaceuticals. We assign a multiplier to each activity/link of the supply chain to represent the fraction of the product that may degrade/be wasted/be lost over the course of that activity. The fraction of lost product depends on the activity since each process of manufacturing, shipment, storage, and distribution results in a different amount of loss. In addition, this fraction need not be the same among various links in the same tier of the supply chain network since different firms and different units of the same firm may experience different amounts of waste, depending on the brand of drug, the efficiency of the technology, the experience of the staff, etc. Also, such multipliers can capture pilferage/theft, a significant issue in drug supply chains.

Although the supply chain network topology in Fig. 5.1 is distinct from the food supply chain network topology in Fig. 4.1, the underlying behavior of the firms is identical since they are competitive profit-maximizers. The representative total cost functions and demand price functions are, as in Chap. 4, with the understanding that the cost functions now are constructed for the pharmaceutical supply chain network links and the demand price functions correspond to the pharmaceutical firms' products at the demand markets. Of course, here, flows refer to pharmaceutical product flows, whereas in Chap. 4, the flows corresponded to fresh produce product flows.

Hence, the notation for the pharmaceutical supply chain network model is similar to the food supply chain network model in Chap. 4. However, the paths connecting the origin/destination pairs of nodes, the firm and demand market pairs, have a different number of links since there are fewer tiers of nodes in the pharmaceutical supply chain network than in the food supply chain network.

The reader should refer to Chap. 4 for the conservation of flow equations. The definitions of the various arc and path multipliers are identical to those in Chaps. 2 through 4, as well. Note that our framework for competitive firms producing perishable products is sufficiently general, with the appropriate topologies defined, to capture sectors of different industries, as the relationships between the pharmaceutical supply chain network in this chapter and the food supply chain network in Chap. 4 demonstrate.

As in Chap. 4, X_i denotes the vector of path flows associated with firm i, where $X_i \equiv \{\{x_p\} | p \in \mathscr{P}^i\} \in R_+^{n_{\mathscr{P}^i}}$, and $\mathscr{P}^i \equiv \cup_{k=1,\ldots,n_R} \mathscr{P}_k^i$. In turn $n_{\mathscr{P}^i}$, denotes the number of paths from firm i to the demand markets. Thus, X is the vector of all the food firms' strategies, that is, $X \equiv \{\{X_i\} | i = 1, \ldots, I\}$.

The profit function of a pharmaceutical firm is defined as the difference between its revenue and its total costs, where the revenue is equal to the summation of the price times the demand at each demand market. The total costs are composed of the total operational costs as well as the total discarding costs of waste on the links in the supply chain network under control by each firm. Hence, the profit function of firm i, denoted by U_i, is expressed here as

$$U_i = \sum_{k=1}^{n_R} \rho_{ik}(d) d_{ik} - \left(\sum_{a \in L^i} \hat{c}_a(f) + \sum_{a \in L^i} \hat{z}_a(f_a) \right). \tag{5.1}$$

Note that the profit function (5.1) is identical to the one in (4.12) since, in both chapters, the firms are profit-maximizers. Here the flows are different and the costs are for pharmaceutical supply chain network activities and the prices are for the different pharmaceutical brands. In lieu of (4.7) and (4.9), the profits of all the firms can be grouped into the I-dimensional vector U:

$$U = U(X). \tag{5.2}$$

In the Cournot–Nash oligopolistic market framework, each firm selects its product path flows in a noncooperative manner, seeking to maximize its own profit, until an equilibrium is achieved, according to Definition 4.1 in Chap. 4.

In other words, an equilibrium is established if no firm can unilaterally improve its profit by changing its production path flows, given the production path flow decisions of the other firms. Here we present an alternative variational inequality formulation in path flows to the one in Theorem 4.1.

Theorem 5.1 (Variational Inequality Formulations). *Assume that for each pharmaceutical firm i; $i = 1,\ldots,I$, the profit function $U_i(X)$ is concave with respect to the variables in X_i and is continuously differentiable. Then $X^* \in K$ is a supply chain generalized network Cournot–Nash equilibrium according to Definition 4.1 if and only if it satisfies the variational inequality*

$$-\sum_{i=1}^{I} \langle \nabla_{X_i} U_i(X^*), X_i - X_i^* \rangle \geq 0, \quad \forall X \in K, \tag{5.3}$$

where $\langle \cdot, \cdot \rangle$ denotes the inner product in the corresponding Euclidean space and $\nabla_{X_i} U_i(X)$ denotes the gradient of $U_i(X)$ with respect to X_i. Variational inequality (5.3), in turn, for the model, is equivalent to the variational inequality: determine the vector of equilibrium path flows and the vector of equilibrium demands $(x^*, d^*) \in K^1$ such that

$$\sum_{i=1}^{I} \sum_{k=1}^{n_R} \sum_{p \in \mathcal{P}_k^i} \left[\frac{\partial \hat{C}_p(x^*)}{\partial x_p} + \frac{\partial \hat{Z}_p(x^*)}{\partial x_p} \right] \times [x_p - x_p^*]$$

$$+ \sum_{i=1}^{I} \sum_{k=1}^{n_R} \left[-\rho_{ik}(d^*) - \sum_{l=1}^{n_R} \frac{\partial \rho_{il}(d^*)}{\partial d_{ik}} d_{il}^* \right] \times [d_{ik} - d_{ik}^*] \geq 0, \quad \forall (x,d) \in K^1, \tag{5.4}$$

where $K^1 \equiv \{(x,d) | x \in R_+^{n_\mathcal{P}}$ and (4.9) holds$\}$, and for notational convenience, as in (4.18), for each path p; $p \in \mathcal{P}_k^i$; $i = 1,\ldots,I$; $k = 1,\ldots,n_R$,:

$$\frac{\partial \hat{C}_p(x)}{\partial x_p} \equiv \frac{\partial \sum_{q \in \mathcal{P}^i} \hat{C}_q(x)}{\partial x_p} = \sum_{b \in L^i} \sum_{a \in L^i} \frac{\partial \hat{c}_b(f)}{\partial f_a} \alpha_{ap},$$

$$\frac{\partial \hat{Z}_p(x)}{\partial x_p} \equiv \frac{\partial \sum_{q \in \mathcal{P}^i} \hat{Z}_q(x)}{\partial x_p} = \sum_{a \in L^i} \frac{\partial \hat{z}_a(f_a)}{\partial f_a} \alpha_{ap}. \tag{5.5}$$

5.2 The Pharmaceutical Supply Chain Network Oligopoly Model

Variational inequality (5.4) can also be reexpressed in terms of link flows as: determine the vector of equilibrium link flows and the vector of equilibrium demands $(f^*, d^*) \in K^2$, *such that*

$$\sum_{i=1}^{I}\sum_{a\in L^i}\left[\sum_{b\in L^i}\frac{\partial \hat{c}_b(f^*)}{\partial f_a} + \frac{\partial \hat{z}_a(f_a^*)}{\partial f_a}\right] \times [f_a - f_a^*]$$
$$+ \sum_{i=1}^{I}\sum_{k=1}^{n_R}\left[-\rho_{ik}(d^*) - \sum_{l=1}^{n_R}\frac{\partial \rho_{il}(d^*)}{\partial d_{ik}}d_{il}^*\right] \times [d_{ik} - d_{ik}^*] \geq 0, \quad \forall (f,d) \in K^2, \quad (5.6)$$

where $K^2 \equiv \{(f,d) | \exists x \geq 0,\ and\ (4.7)\ and\ (4.9)\ hold\}$.

Proof. Variational inequality (5.3) follows directly from the first part of the proof of Theorem 4.1. Note that, as before,

$$\nabla_{X_i} U_i(X) = \left[\frac{\partial U_i}{\partial x_p}; p \in \mathscr{P}_k^i; k = 1, \ldots, n_R\right], \quad (5.7)$$

where now, for each path $p;\ p \in \mathscr{P}_k^i$, we have

$$\frac{\partial U_i}{\partial x_p} = \frac{\partial \left[\sum_{l=1}^{n_R}\rho_{il}(d)d_{il} - \sum_{b\in L^i}\hat{c}_b(f) - \sum_{b\in L^i}\hat{z}_b(f_b)\right]}{\partial x_p}$$
$$= \sum_{l=1}^{n_R}\frac{\partial [\rho_{il}(d)d_{il}]}{\partial x_p} - \frac{\partial \left[\sum_{b\in L^i}\hat{c}_b(f)\right]}{\partial x_p} - \frac{\partial \left[\sum_{b\in L^i}\hat{z}_b(f_b)\right]}{\partial x_p}$$
$$= \rho_{ik}(d)\mu_p + \sum_{l=1}^{n_R}\frac{\partial \rho_{il}(d)}{\partial d_{ik}}\frac{\partial d_{ik}}{\partial x_p}d_{il} - \sum_{a\in L^i}\frac{\partial \left[\sum_{b\in L^i}\hat{c}_b(f)\right]}{\partial f_a}\frac{\partial f_a}{\partial x_p}$$
$$- \sum_{a\in L^i}\frac{\partial \left[\sum_{b\in L^i}\hat{z}_b(f_b)\right]}{\partial f_a}\frac{\partial f_a}{\partial x_p}$$
$$= \rho_{ik}(d)\mu_p + \sum_{l=1}^{n_R}\frac{\partial \rho_{il}(d)}{\partial d_{ik}}\mu_p d_{il} - \sum_{a\in L^i}\sum_{b\in L^i}\frac{\partial \hat{c}_b(f)}{\partial f_a}\alpha_{ap} - \sum_{a\in L^i}\frac{\partial \hat{z}_a(f_a)}{\partial f_a}\alpha_{ap}. \quad (5.8)$$

Multiplying the expression in (5.8) by a minus sign and by the term $(x_p - x_p^*)$ and summing up over all paths p and making use of the definition of the feasible set K^1, with notice to constraint (4.9), and recalling the definitions of $\frac{\partial \hat{C}_p(x)}{\partial x_p}$ and $\frac{\partial \hat{Z}_p(x)}{\partial x_p}$ in (5.5)—the equivalence of which is established in Chap. 2—after algebraic simplification, yield variational inequality (5.4). By using then Eq. (4.7), variational inequality (5.6) follows from (5.4). □

Existence and uniqueness results are as discussed in Chap. 4.

The above model is now related to several models in this book. If the arc multipliers are all equal to 1, then the model is related to the sustainable fashion supply chain network model that will be discussed in Chap. 6. In that model, however, another criterion, in addition to the profit maximization one, is emission minimization.

If the product is homogeneous, that is, the consumers are indifferent as to who is the producer, and all the arc multipliers are equal to 1, and the total costs are separable, then the above model collapses to the supply chain network oligopoly model of Nagurney (2010) in which synergies associated with mergers and acquisitions were assessed.

In addition, if there is only a single organization/firm, there is no product differentiation, and the demands are subject to uncertainty, with the inclusion of expected costs due to shortages or excess supplies, the total operational cost functions are separable, and a criterion of risk is added, then the model above is related to the blood supply chain network optimization model in Chap. 2.

We now present a simple numerical example in order to illustrate the model.

5.2.1 An Illustrative Pharmaceutical Supply Chain Network Example

In this example, two pharmaceutical firms compete in a duopoly with a single demand market (see Fig. 5.2). The two firms produce differentiated, but substitutable, brand drugs 1 and 2, corresponding to Firm 1 and Firm 2, respectively.

The total cost functions on the various links of manufacturing, shipment, storage, and distribution are, for simplicity, separable, that is, the total cost function on a link depends only on the flow on that link:

$$\hat{c}_1(f_1) = 5f_1^2 + 8f_1, \hat{c}_2(f_2) = 2f_2^2 + f_2, \hat{c}_3(f_3) = 3f_3^2 + 4f_3, \hat{c}_4(f_4) = 2f_4^2 + 5f_4,$$

$$\hat{c}_5(f_5) = 7f_5^2 + 3f_5, \hat{c}_6(f_6) = 2f_6^2 + 2f_6, \hat{c}_7(f_7) = 3.5f_7^2 + f_7, \hat{c}_8(f_8) = 1.5f_8^2 + 4f_8.$$

The arc multipliers are

$$\alpha_1 = .95, \alpha_2 = \alpha_3 = .99, \alpha_4 = 1.00, \alpha_5 = .98, \alpha_6 = 1.00, \alpha_7 = .97, \alpha_8 = 1.00.$$

The total discarding cost functions on the links are assumed to be identical, that is,

$$\hat{z}_a(f_a) = .5f_a^2, \quad \forall a \in L.$$

The firms compete in the demand market R_1, and the consumers reveal their preferences for the two products through the following nonseparable demand price functions:

$$\rho_{11}(d) = -3d_{11} - d_{21} + 200, \quad \rho_{21}(d) = -4d_{21} - 1.5d_{11} + 300.$$

In this supply chain network, there exists one path corresponding to each firm, denoted by $p_1 = (1,2,3,4)$ and $p_2 = (5,6,7,8)$. Thus, variational inequality (5.4) can, for this example, since $d_{11}^* = x_{p_1}^* \mu_{p_1}$ and $d_{21}^* = x_{p_2}^* \mu_{p_2}$, and $d_{11} = x_{p_1} \mu_{p_1}$ and $d_{21} = x_{p_2} \mu_{p_2}$, be rewritten as

5.2 The Pharmaceutical Supply Chain Network Oligopoly Model

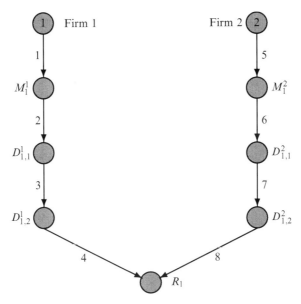

Fig. 5.2 Supply chain network topology for the pharmaceutical duopoly in the illustrative example

$$\left[\frac{\partial \hat{C}_{p_1}(x^*)}{\partial x_{p_1}} + \frac{\partial \hat{Z}_{p_1}(x^*)}{\partial x_{p_1}} - \rho_{11}(d^*)\mu_{p_1} - \frac{\partial \rho_{11}(d^*)}{\partial d_{11}}\mu_{p_1} \times d_{11}^*\right] \times [x_{p_1} - x_{p_1}^*]$$
$$+ \left[\frac{\partial \hat{C}_{p_2}(x^*)}{\partial x_{p_2}} + \frac{\partial \hat{Z}_{p_2}(x^*)}{\partial x_{p_2}} - \rho_{21}(d^*)\mu_{p_2} - \frac{\partial \rho_{21}(d^*)}{\partial d_{21}}\mu_{p_2} \times d_{21}^*\right] \times [x_{p_2} - x_{p_2}^*] \geq 0,$$
$$\forall x \in R_+^2. \quad (5.9)$$

Under the assumption that $x_{p_1}^* > 0$ and $x_{p_2}^* > 0$, the two expressions on the left-hand side of inequality (5.9) must be equal to zero, that is

$$\left[\frac{\partial \hat{C}_{p_1}(x^*)}{\partial x_{p_1}} + \frac{\partial \hat{Z}_{p_1}(x^*)}{\partial x_{p_1}} - \rho_{11}(d^*)\mu_{p_1} - \frac{\partial \rho_{11}(d^*)}{\partial d_{11}}\mu_{p_1} \times d_{11}^*\right] \times [x_{p_1} - x_{p_1}^*] = 0,$$
(5.10a)

and

$$\left[\frac{\partial \hat{C}_{p_2}(x^*)}{\partial x_{p_2}} + \frac{\partial \hat{Z}_{p_2}(x^*)}{\partial x_{p_2}} - \rho_{21}(d^*)\mu_{p_2} - \frac{\partial \rho_{21}(d^*)}{\partial d_{21}}\mu_{p_2} \times d_{21}^*\right] \times [x_{p_2} - x_{p_2}^*] = 0.$$
(5.10b)

Since each of the paths flows must be nonnegative, we know that the term preceding the multiplication sign in both (5.10a) and (5.10b) must be equal to zero.

Calculating the values of the multipliers and then substituting those values, as well as the given functions for this example into (5.5), we can determine the partial derivatives of the total operational cost and the total discarding cost functions in (5.10a) and (5.10b). Furthermore, the partial derivatives of the given demand price functions can be calculated and substituted into the above. The path multipliers are equal to

$$\mu_{p_1} = \alpha_1 \times \alpha_2 \times \alpha_3 \times \alpha_4 = .95 \times .99 \times .99 \times 1 = .93,$$
$$\mu_{p_2} = \alpha_5 \times \alpha_6 \times \alpha_7 \times \alpha_8 = .98 \times 1 \times .97 \times 1 = .95.$$

Arithmetic calculations, using the above substitutions, yield the system of equations:

$$31.24 x^*_{p_1} + 0.89 x^*_{p_2} = 168.85,$$
$$1.33 x^*_{p_1} + 38.33 x^*_{p_2} = 274.46. \tag{5.11}$$

Thus, the equilibrium solution corresponding to the path flow of brand drugs produced by Firms 1 and 2 is

$$x^*_{p_1} = 5.21, \quad x^*_{p_2} = 6.98.$$

Using (4.7), the equilibrium link flows can be calculated as

$$f^*_1 = 5.21, \quad f^*_2 = 4.95, \quad f^*_3 = 4.90, \quad f^*_4 = 4.85,$$
$$f^*_5 = 6.98, \quad f^*_6 = 6.84, \quad f^*_7 = 6.84, \quad f^*_8 = 6.64.$$

Using (4.9), the equilibrium demands for products of the two pharmaceutical firms are

$$d^*_{11} = 4.85, \quad d^*_{21} = 6.64.$$

Hence, the incurred equilibrium prices of the two branded drugs are

$$\rho_{11}(d^*) = 178.82, \quad \rho_{21}(d^*) = 266.19.$$

Note that even though the price of Firm 2's product is calculated to be higher, the market has a slightly stronger tendency toward this product as opposed to the product of Firm 1. This is due to the willingness of the consumers to spend more on one product which can be a consequence of the reputation, the perceived quality, etc., of Firm 2's brand drug. The profits of the two firms are

$$U_1(X^*) = 713.52, \quad U_2(X^*) = 1,537.32.$$

Next, we discuss a special case of our model in which the pharmaceutical firms produce a homogeneous drug.

Corollary 5.1. *Assume that the pharmaceutical firms produce a homogeneous drug. We may then denote the demand for the homogeneous drug and its demand price at demand market R_k, respectively, by d_k and ρ_k, instead of by d_{ik} and ρ_{ik}. Consequently, the following equation, which replaces (4.9), must then hold:*

$$\sum_{i=1}^{I} \sum_{p \in \mathscr{P}^i_k} x_p \mu_p = d_k, \quad k = 1, \ldots, n_R. \tag{5.12}$$

5.2 The Pharmaceutical Supply Chain Network Oligopoly Model

Then, the profit function (5.1) can be rewritten as

$$U_i = \sum_{k=1}^{n_R} \rho_k(d) \sum_{p \in \mathscr{P}_k^i} \mu_p x_p - \left(\sum_{a \in L^i} \hat{c}_a(f) + \sum_{a \in L^i} \hat{z}_a(f_a) \right). \tag{5.13}$$

The corresponding variational inequality (5.4) in terms of path flows can be rewritten as determine $(x^*, d^*) \in K^3$ *such that*

$$\sum_{i=1}^{I} \sum_{k=1}^{n_R} \sum_{p \in \mathscr{P}_k^i} \left[\frac{\partial \hat{C}_p(x^*)}{\partial x_p} + \frac{\partial \hat{Z}_p(x^*)}{\partial x_p} - \sum_{l=1}^{n_R} \frac{\partial \rho_l(d^*)}{\partial d_k} \mu_p \sum_{q \in \mathscr{P}_l^i} \mu_q x_q^* \right] \times [x_p - x_p^*]$$

$$+ \sum_{k=1}^{n_R} [-\rho_k(d^*)] \times [d_k - d_k^*] \geq 0, \quad \forall (x, d) \in K^3, \tag{5.14}$$

where $K^3 \equiv \{(x,d) | x \in R_+^{n_{\mathscr{P}}} \text{ and } (5.12) \text{ holds}\}$.

Proof. According to the proof of Theorem 5.1, variational inequality (5.14) can be proved by replacing d_{ik} and ρ_{ik}, respectively, by d_k and ρ_k. □

It is interesting to note that our supply chain generalized network oligopoly model can also capture the competition in the pharmaceutical industry even when the demands d_{ik} are fixed, for all brands i, at all demand markets R_k, since we consider total cost functions that are not separable. Fixed demands for pharmaceutical products arise, for example, in the case of certain hospital and medical procedures, which need to be scheduled in advance. For example, the supply chain for medical nuclear products, as discussed in Chap. 3, is characterized by fixed demands since medical procedures that use radioisotopes need to be scheduled in advance. Moreover, radioisotopes, since they are subject to radioactive decay, are not only perishable but also time-sensitive.

Corollary 5.2. *Assume that the demand d_{ik} for firm i's pharmaceutical; $i = 1, \ldots, I$, at demand market R_k; $k = 1, \ldots, n_R$, is fixed. The demand price of firm i's product at demand market R_k will then also be fixed; we denote this price by $\bar{\rho}_{ik}$. The profit function (5.1) can then be rewritten as*

$$U_i = \sum_{k=1}^{n_R} \bar{\rho}_{ik} d_{ik} - \left(\sum_{a \in L^i} \hat{c}_a(f) + \sum_{a \in L^i} \hat{z}_a(f_a) \right), \tag{5.15}$$

where the revenue of firm i, $\sum_{k=1}^{n_R} \bar{\rho}_{ik} d_{ik}$, is fixed. Therefore, the corresponding variational inequality (5.4) in terms of path flows simplifies, in this case, to: determine $x^* \in K^4$ such that

$$\sum_{i=1}^{I} \sum_{k=1}^{n_R} \sum_{p \in \mathscr{P}_k^i} \left[\frac{\partial \hat{C}_p(x^*)}{\partial x_p} + \frac{\partial \hat{Z}_p(x^*)}{\partial x_p} \right] \times [x_p - x_p^*] \geq 0, \quad \forall x \in K^4, \tag{5.16}$$

where $K^4 \equiv \{x|x \geq 0, \text{ and } (4.9) \text{ is satisfied with the } d_{ik}\text{s known and fixed}, \forall i, \forall k\}$.

Similarly, variational inequality (5.16) can be reexpressed in terms of link flows as: determine $f^* \in K^5$, such that

$$\sum_{i=1}^{I} \sum_{a \in L^i} \left[\sum_{b \in L^i} \frac{\partial \hat{c}_b(f^*)}{\partial f_a} + \frac{\partial \hat{z}_a(f_a^*)}{\partial f_a} \right] \times [f_a - f_a^*] \geq 0, \quad \forall f \in K^5, \qquad (5.17)$$

where $K^5 \equiv \{f|\exists x \geq 0, \text{ and } (4.7) \text{ and } (4.9) \text{ hold with the } d_{ik}\text{s known and fixed}, \forall i, \forall k\}$.

Proof. Based on the proof of Theorem 5.1, variational inequality (5.16) can be proved by eliminating the corresponding term of firm i's revenue in the profit function, since the revenue of firm i, $\sum_{k=1}^{n_R} \bar{\rho}_{ik} d_{ik}$, is fixed. Also, using Eq. (4.7), variational inequality (5.17) follows from (5.16). □

We further discuss the specific case in which the pharmaceutical companies produce a homogeneous drug and the demand at each demand market is fixed.

Corollary 5.3. *Assume that the pharmaceutical firms produce a homogeneous drug. We may then denote the demand for the homogeneous drug and its demand price at demand market R_k, respectively, by d_k and $\bar{\rho}_k$, instead of by d_{ik} and ρ_{ik}. Consequently, the following equation, which replaces (4.9), must then hold:*

$$\sum_{i=1}^{I} \sum_{p \in \mathcal{P}_k^i} x_p \mu_p = d_k, \quad k = 1, \ldots, n_R. \qquad (5.18)$$

Assume also that the demand d_k at demand market R_k; $k = 1, \ldots, n_R$ is fixed, as well as the demand price $\bar{\rho}_k$. Then, the profit function (5.1) for firm i can be rewritten as

$$U_i = \sum_{k=1}^{n_R} \bar{\rho}_k \sum_{p \in \mathcal{P}_k^i} \mu_p x_p - \left(\sum_{a \in L^i} \hat{c}_a(f) + \sum_{a \in L^i} \hat{z}_a(f_a) \right). \qquad (5.19)$$

The corresponding variational inequality in terms of path flows is: determine $x^* \in K^6$ such that

$$\sum_{i=1}^{I} \sum_{k=1}^{n_R} \sum_{p \in \mathcal{P}_k^i} \left[\frac{\partial \hat{C}_p(x^*)}{\partial x_p} + \frac{\partial \hat{Z}_p(x^*)}{\partial x_p} \right] \times [x_p - x_p^*] \geq 0, \quad \forall x \in K^6, \qquad (5.20)$$

where $K^6 \equiv \{x|x \geq 0, \text{ and } (5.18) \text{ is satisfied with the } d_k\text{s known and fixed}, \forall k\}$.

Similarly, the corresponding variational inequality in terms of link flows is: determine $f^* \in K^7$, such that

$$\sum_{i=1}^{I} \sum_{a \in L^i} \left[\sum_{b \in L^i} \frac{\partial \hat{c}_b(f^*)}{\partial f_a} + \frac{\partial \hat{z}_a(f_a^*)}{\partial f_a} \right] \times [f_a - f_a^*] \geq 0, \quad \forall f \in K^7, \qquad (5.21)$$

where $K^7 \equiv \{f|\exists x \geq 0, \text{ and } (4.7) \text{ and } (5.18) \text{ hold with the } d_k\text{s known and fixed}, \forall k\}$.

5.2 The Pharmaceutical Supply Chain Network Oligopoly Model

Proof. Following the proof of Theorem 5.1, we have

$$\sum_{i=1}^{I}\sum_{k=1}^{n_R}\sum_{p\in\mathscr{P}_k^i}\left[\frac{\partial \hat{C}_p(x^*)}{\partial x_p}+\frac{\partial \hat{Z}_p(x^*)}{\partial x_p}-\bar{\rho}_k\mu_p\right]\times[x_p-x_p^*]\geq 0, \quad \forall x\in K^6, \quad (5.22)$$

which is equivalent to

$$\sum_{i=1}^{I}\sum_{k=1}^{n_R}\sum_{p\in\mathscr{P}_k^i}\left[\frac{\partial \hat{C}_p(x^*)}{\partial x_p}+\frac{\partial \hat{Z}_p(x^*)}{\partial x_p}\right]\times[x_p-x_p^*]$$
$$-\sum_{k=1}^{n_R}\bar{\rho}_k\left[\sum_{i=1}^{I}\sum_{p\in\mathscr{P}_k^i}\mu_p x_p-\sum_{i=1}^{I}\sum_{p\in\mathscr{P}_k^i}\mu_p x_p^*\right]\geq 0, \quad \forall x\in K^6. \quad (5.23)$$

Applying now Eq. (5.18) to (5.23) yields variational inequality (5.20). Also, using Eq. (4.7), variational inequality (5.21) then follows from (5.20). □

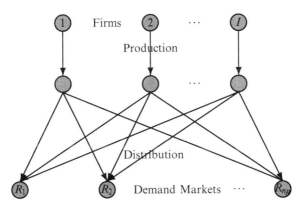

Fig. 5.3 The *Bipartite* structure of a special case of the supply chain problem

Remark. It is worth noting the relationship of the supply chain network model in this chapter (as well as the one in Chap. 4) to spatial oligopolistic market equilibrium models (cf. Dafermos and Nagurney 1987). The latter can be viewed as precursors to supply chain network game theory models. For example, if we assume a homogeneous product, aggregate each firm's production, processing, and storage activities, and assume only a single mode of transportation from each firm to each demand market, then the network in Fig. 5.1 (also that in Fig. 4.1) reduces to the supply chain network in Fig. 5.3. The inclusion of perishability provides a nice extension to such models, as does the product/brand differentiation. Nagurney and Li (2012) used such a topology to formulate a dynamic oligopoly model with quality competition, whereas Nagurney and Yu (2012) formulated time-based competition with delivery times as strategic variables on such networks.

5.3 The Algorithm

As in Chap. 4, we utilize the Euler method as given in expression (2.59) for our numerical examples in our case study, but here we provide an alternative statement of the closed form expressions.

5.3.1 Explicit Formulae for the Euler Method Applied to the Pharmaceutical Supply Chain Generalized Network Oligopoly Variational Inequality (5.4)

The elegance of this procedure for the computation of solutions to our pharmaceutical supply chain generalized network oligopoly model with product differentiation governed by variational inequality (5.4) can be seen in the following explicit formulae. In particular, we have the closed form expression for the flow on each path $p \in \mathscr{P}_k^i; i = 1, \ldots, I; k = 1, \ldots, n_R$, at iteration $\tau + 1$:

$$x_p^{\tau+1} = \max\{0, x_p^\tau + a_\tau(\rho_{ik}(d^\tau)\mu_p + \sum_{l=1}^{n_R} \frac{\partial \rho_{il}(d^\tau)}{\partial d_{ik}} \mu_p d_{il}^\tau - \frac{\partial \hat{C}_p(x^\tau)}{\partial x_p} - \frac{\partial \hat{Z}_p(x^\tau)}{\partial x_p})\}, \tag{5.24a}$$

with the demands being updated according to

$$d_{ik}^{\tau+1} = \sum_{p \in \mathscr{P}_k^i} \mu_p x_p^{\tau+1}. \tag{5.24b}$$

In the next section, we solve several cases of pharmaceutical supply chain network problems using the above algorithmic scheme.

Furthermore, we emphasize that one can also utilize the Euler method to solve the supply chain generalized network models as in Corollaries 5.1, 5.2, and 5.3, with appropriate adaptations.

5.4 A Case Study

In this case study, we apply the Euler method to solve a set of pharmaceutical supply chain network oligopoly problems. The examples focus on cholesterol-regulating drug competition in the USA under various scenarios, including the expiration of the patent rights to a popular brand and the emergence of its generic substitute. Although the scenarios are stylized, they illustrate the modeling and algorithmic framework described in this chapter. For purposes of transparency and reproducibility, we provide both the input and the output data.

5.4 A Case Study

Scenario 1. This scenario is assumed to occur in the third quarter of 2011 prior to the expiration of the patent for Lipitor. Lipitor was once the top-selling pharmaceutical brand in the world with more than $5 billion of sales in the USA alone in 2011 (Rossi 2011). Firm 1 represents a multinational pharmaceutical firm that holds the patent. Firm 2 is also one of the largest global pharmaceutical companies that has been producing another cholesterol-regulating drug whose patent expired in 2006. Each of these firms is assumed to have two manufacturing units and three storage/distribution centers, as illustrated in Fig. 5.4.

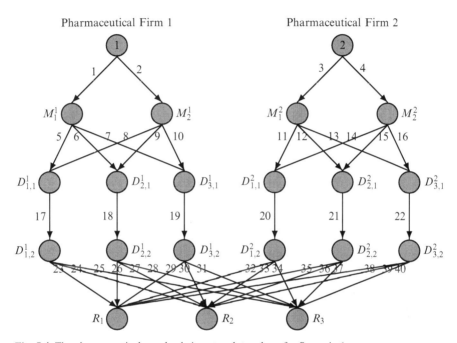

Fig. 5.4 The pharmaceutical supply chain network topology for Scenario 1

R_1, R_2, and R_3 represent three pharmacies located in the USA. The demand price functions corresponding to these three demand markets for each of the two brands were

$$\rho_{11}(d) = -1.1d_{11} - 0.9d_{21} + 275, \quad \rho_{21}(d) = -1.2d_{21} - 0.7d_{11} + 210,$$
$$\rho_{12}(d) = -0.9d_{12} - 0.8d_{22} + 255, \quad \rho_{22}(d) = -1.0d_{22} - 0.5d_{12} + 200,$$
$$\rho_{13}(d) = -1.4d_{13} - 1.0d_{23} + 265, \quad \rho_{23}(d) = -1.5d_{23} - 0.4d_{13} + 186.$$

The demand for each firm represents the number of packages of an equivalent dosage for each brand over a time period of a week.

The arc multipliers, the total operational cost functions, and the total discarding cost functions are reported in Table 5.1. These cost functions have been selected based on the average values of the prices, the shipping costs, etc., available on the web. The values of arc multipliers, although hypothetical, are constructed in order to reflect the percentage of perishability/waste/loss associated with the various activities in these supply chains.

The Euler method (cf. (5.24a) and (5.24b)) for the solution of variational inequality (5.4) was implemented in MATLAB. The sequence $\{a_\tau\} = .1(1, \frac{1}{2}, \frac{1}{2}, \ldots)$, and the convergence tolerance was 10^{-6}. In other words, the absolute value of the difference between each path flow in two consecutive iterations was less than or equal to this tolerance. We initialized the algorithm by setting the path flows equal to 10. Table 5.1 provides the computed equilibrium link product flows. We report the equilibrium link flows, rather than the path flows, due to space limitation.

Table 5.1 Link multipliers, total operational cost, and total discarding cost functions and equilibrium link flow solution for Scenario 1

Link a	α_a	$\hat{c}_a(f_a)$	$\hat{z}_a(f_a)$	f_a^*	Link a	α_a	$\hat{c}_a(f_a)$	$\hat{z}_a(f_a)$	f_a^*
1	.95	$5f_1^2 + 8f_1$	$.5f_1^2$	13.73	21	.98	$2.1f_{21}^2 + 3f_{21}$	$.35f_{21}^2$	5.42
2	.97	$7f_2^2 + 3f_2$	$.4f_2^2$	10.77	22	.99	$1.9f_{22}^2 + 2.5f_{22}$	$.5f_{22}^2$	6.00
3	.96	$6.5f_3^2 + 4f_3$	$.3f_3^2$	8.42	23	1.00	$.5f_{23}^2 + 2f_{23}$	$.6f_{23}^2$	3.56
4	.98	$5f_4^2 + 7f_4$	$.35f_4^2$	10.55	24	1.00	$.7f_{24}^2 + f_{24}$	$.6f_{24}^2$	1.66
5	1.00	$.7f_5^2 + f_5$	$.5f_5^2$	5.21	25	.99	$.5f_{25}^2 + .8f_{25}$	$.6f_{25}^2$	2.82
6	.99	$.9f_6^2 + 2f_6$	$.5f_6^2$	3.36	26	.99	$.6f_{26}^2 + f_{26}$	$.45f_{26}^2$	3.34
7	1.00	$.5f_7^2 + f_7$	$.5f_7^2$	4.47	27	.99	$.7f_{27}^2 + .8f_{27}$	$.4f_{27}^2$	1.24
8	.99	$f_8^2 + 2f_8$	$.6f_8^2$	3.02	28	.98	$.4f_{28}^2 + .8f_{28}$	$.45f_{28}^2$	2.59
9	1.00	$.7f_9^2 + 3f_9$	$.6f_9^2$	3.92	29	1.00	$.3f_{29}^2 + 3f_{29}$	$.55f_{29}^2$	3.45
10	1.00	$.6f_{10}^2 + 1.5f_{10}$	$.6f_{10}^2$	3.50	30	1.00	$.75f_{30}^2 + f_{30}$	$.55f_{30}^2$	1.28
11	.99	$.8f_{11}^2 + 2f_{11}$	$.4f_{11}^2$	3.10	31	1.00	$.65f_{31}^2 + f_{31}$	$.55f_{31}^2$	3.09
12	.99	$.8f_{12}^2 + 5f_{12}$	$.4f_{12}^2$	2.36	32	.99	$.5f_{32}^2 + 2f_{32}$	$.3f_{32}^2$	2.54
13	.98	$.9f_{13}^2 + 4f_{13}$	$.4f_{13}^2$	2.63	33	.99	$.4f_{33}^2 + 3f_{33}$	$.3f_{33}^2$	3.43
14	1.00	$.8f_{14}^2 + 2f_{14}$	$.5f_{14}^2$	3.79	34	1.00	$.5f_{34}^2 + 3.5f_{34}$	$.4f_{34}^2$	0.75
15	.99	$.9f_{15}^2 + 3f_{15}$	$.5f_{15}^2$	3.12	35	.98	$.4f_{35}^2 + 2f_{35}$	$.55f_{35}^2$	1.72
16	1.00	$1.1f_{16}^2 + 3f_{16}$	$.6f_{16}^2$	3.43	36	.98	$.3f_{36}^2 + 2.5f_{36}$	$.55f_{36}^2$	2.64
17	.98	$2f_{17}^2 + 3f_{17}$	$.45f_{17}^2$	8.20	37	.99	$.55f_{37}^2 + 2f_{37}$	$.55f_{37}^2$	0.95
18	.99	$2.5f_{18}^2 + f_{18}$	$.55f_{18}^2$	7.25	38	1.00	$.35f_{38}^2 + 2f_{38}$	$.4f_{38}^2$	3.47
19	.98	$2.4f_{19}^2 + 1.5f_{19}$	$.5f_{19}^2$	7.97	39	1.00	$.4f_{39}^2 + 5f_{39}$	$.4f_{39}^2$	2.47
20	.98	$1.8f_{20}^2 + 3f_{20}$	$.3f_{20}^2$	6.85	40	.98	$.55f_{40}^2 + 2f_{40}$	$.6f_{40}^2$	0.00

The values of the equilibrium link flows in Table 5.1 demonstrate the impact of perishability of the product on the supply chain network links of each pharmaceutical firm. Under the above demand price functions, the computed equilibrium demands for each of the two brands were

$$d_{11}^* = 10.32, \quad d_{21}^* = 7.66,$$
$$d_{12}^* = 4.17, \quad d_{22}^* = 8.46,$$
$$d_{13}^* = 8.41, \quad d_{23}^* = 1.69.$$

5.4 A Case Study

The incurred equilibrium prices were

$$\rho_{11}(d^*) = 256.75, \quad \rho_{21}(d^*) = 193.58,$$
$$\rho_{12}(d^*) = 244.48, \quad \rho_{22}(d^*) = 189.46,$$
$$\rho_{13}(d^*) = 251.52, \quad \rho_{23}(d^*) = 180.09.$$

Note that Firm 1, which produces the top-selling product, captures the majority of the market share at demand markets 1 and 3, despite its higher prices. While this firm has a slight advantage over its competitor in demand market 1, it has almost entirely seized demand market 3. Consequently, several links connecting Firm 2 to demand market 3 have insignificant flows including link 40 with a flow equal to zero. In contrast, Firm 2 dominates demand market 2, due to the consumers' willingness to lean toward this product there, perhaps as a consequence of the lower price, the perception of quality, etc. The profits of the two firms were:

$$U_1(X^*) = 2{,}936.52, \quad U_2(X^*) = 1{,}675.89.$$

Firm 1 still holds the patent to its branded drug and, thus, makes a higher profit from selling cholesterol-regulating drugs. In contrast, Firm 2 has completed the competition-free timespan for its branded drug a few years ago, due to the expiration of its patent, allowing the manufacture of generic competitors. Hence, fewer numbers of consumers choose this product as compared to the product of Firm 1 leading to a higher profit for Firm 1.

The next scenario explores the situation of the cholesterol-lowering drug market in the first quarter of 2012, when the patent of Firm 1's product has just expired as well and a third firm steps up to produce a generic substitute of this product.

Scenario 2. The patent for Firm 1's drug has expired despite all the legal and political efforts to extend the patent. A manufacturer of generic drugs, Firm 3, introduces a generic substitute by reproducing the active ingredients. Firm 3 is assumed to have two manufacturing plants, two distribution centers, as well as two storage facilities to supply the same three demand markets as in Scenario 1. The supply chain network topology for this scenario is illustrated in Fig. 5.5.

In this scenario the new generic drug has just been released. We assume that the demand price functions for the products of Firms 1 and 2 will remain the same as in Scenario 1. The demand price functions corresponding to the product of Firm 3 for demand markets 1, 2, and 3 are as follows:

$$\rho_{31}(d) = -0.9d_{31} - 0.6d_{11} - 0.8d_{21} + 150,$$
$$\rho_{32}(d) = -0.8d_{32} - 0.5d_{12} - 0.6d_{22} + 130,$$
$$\rho_{33}(d) = -0.9d_{33} - 0.7d_{13} - 0.5d_{23} + 133.$$

Table 5.2 displays the arc multipliers, the total operational and the total discarding cost functions for all links in Fig. 5.5 and the computed values of the equilibrium link flows.

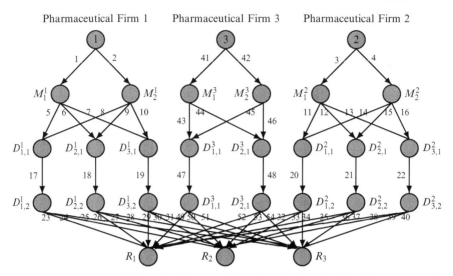

Fig. 5.5 The pharmaceutical supply chain network topology for Scenarios 2 and 3

The equilibrium product flows of Firms 1 and 2 on links 1 through 40 are identical to the corresponding values in Scenario 1. When the new product produced by Firm 3 is just introduced, the manufacturers of the two existing products do not experience an immediate impact on their respective demands of branded drugs. Consequently, the computed equilibrium demands for the products of Firms 1 and 2 at the demand markets will remain as in Scenario 1. However, the equilibrium demands for the product of Firm 3 are

$$d_{31}^* = 5.17, \quad d_{32}^* = 3.18, \quad d_{33}^* = 3.01.$$

Under the above assumptions, the equilibrium prices associated with the branded drugs 1 and 2 at the demand markets remain the same, whereas the incurred equilibrium prices of generic drug 3 were

$$\rho_{31}(d^*) = 133.02, \quad \rho_{32}(d^*) = 120.30, \quad \rho_{33}(d^*) = 123.55.$$

These are significantly lower than the respective prices of its competitors in all the demand markets. Thus, the profit that Firm 3 derives from manufacturing and delivering the new generic drug was

$$U_3(X^*) = 637.38,$$

while the profits of Firms 1 and 2 remain unchanged. In the next scenario, we investigate the situation when the consumers become aware of the new generic substitute.

Scenario 3. In this scenario, we consider the case of the cholesterol drug market at some later time. The generic product of Firm 3 has now become well established and

5.4 A Case Study

Table 5.2 Link multipliers, total operational cost, and total discarding cost functions and equilibrium link flow solution for Scenario 2

Link a	α_a	$\hat{c}_a(f_a)$	$\hat{z}_a(f_a)$	f_a^*	Link a	α_a	$\hat{c}_a(f_a)$	$\hat{z}_a(f_a)$	f_a^*
1	.95	$5f_1^2+8f_1$	$.5f_1^2$	13.73	28	.98	$.4f_{28}^2+.8f_{28}$	$.45f_{28}^2$	2.59
2	.97	$7f_2^2+3f_2$	$.4f_2^2$	10.77	29	1.00	$.3f_{29}^2+3f_{29}$	$.55f_{29}^2$	3.45
3	.96	$6.5f_3^2+4f_3$	$.3f_3^2$	8.42	30	1.00	$.75f_{30}^2+f_{30}$	$.55f_{30}^2$	1.28
4	.98	$5f_4^2+7f_4$	$.35f_4^2$	10.55	31	1.00	$.65f_{31}^2+f_{31}$	$.55f_{31}^2$	3.09
5	1.00	$.7f_5^2+f_5$	$.5f_5^2$	5.21	32	.99	$.5f_{32}^2+2f_{32}$	$.3f_{32}^2$	2.54
6	.99	$.9f_6^2+2f_6$	$.5f_6^2$	3.36	33	.99	$.4f_{33}^2+3f_{33}$	$.3f_{33}^2$	3.43
7	1.00	$.5f_7^2+f_7$	$.5f_7^2$	4.47	34	1.00	$.5f_{34}^2+3.5f_{34}$	$.4f_{34}^2$	0.75
8	.99	$f_8^2+2f_8$	$.6f_8^2$	3.02	35	.98	$.4f_{35}^2+2f_{35}$	$.55f_{35}^2$	1.72
9	1.00	$.7f_9^2+3f_9$	$.6f_9^2$	3.92	36	.98	$.3f_{36}^2+2.5f_{36}$	$.55f_{36}^2$	2.64
10	1.00	$.6f_{10}^2+1.5f_{10}$	$.6f_{10}^2$	3.50	37	.99	$.55f_{37}^2+2f_{37}$	$.55f_{37}^2$	0.95
11	.99	$.8f_{11}^2+2f_{11}$	$.4f_{11}^2$	3.10	38	1.00	$.35f_{38}^2+2f_{38}$	$.4f_{38}^2$	3.47
12	.99	$.8f_{12}^2+5f_{12}$	$.4f_{12}^2$	2.36	39	1.00	$.4f_{39}^2+5f_{39}$	$.4f_{39}^2$	2.47
13	.98	$.9f_{13}^2+4f_{13}$	$.4f_{13}^2$	2.63	40	.98	$.55f_{40}^2+2f_{40}$	$.6f_{40}^2$	0.00
14	1.00	$.8f_{14}^2+2f_{14}$	$.5f_{14}^2$	3.79	41	.97	$3f_{41}^2+12f_{41}$	$.3f_{41}^2$	6.17
15	.99	$.9f_{15}^2+3f_{15}$	$.5f_{15}^2$	3.12	42	.96	$2.7f_{42}^2+10f_{42}$	$.4f_{42}^2$	6.23
16	1.00	$1.1f_{16}^2+3f_{16}$	$.6f_{16}^2$	3.43	43	.98	$1.1f_{43}^2+6f_{43}$	$.45f_{43}^2$	3.23
17	.98	$2f_{17}^2+3f_{17}$	$.45f_{17}^2$	8.20	44	.98	$.9f_{44}^2+5f_{44}$	$.45f_{44}^2$	2.75
18	.99	$2.5f_{18}^2+f_{18}$	$.55f_{18}^2$	7.25	45	.97	$1.3f_{45}^2+6f_{45}$	$.5f_{45}^2$	3.60
19	.98	$2.4f_{19}^2+1.5f_{19}$	$.5f_{19}^2$	7.97	46	.99	$1.5f_{46}^2+7f_{46}$	$.55f_{46}^2$	2.38
20	.98	$1.8f_{20}^2+3f_{20}$	$.3f_{20}^2$	6.85	47	.98	$1.5f_{47}^2+4f_{47}$	$.4f_{47}^2$	6.66
21	.98	$2.1f_{21}^2+3f_{21}$	$.35f_{21}^2$	5.42	48	.98	$2.1f_{48}^2+6f_{48}$	$.45f_{48}^2$	5.05
22	.99	$1.9f_{22}^2+2.5f_{22}$	$.5f_{22}^2$	6.00	49	.99	$.6f_{49}^2+3f_{49}$	$.55f_{49}^2$	3.79
23	1.00	$.5f_{23}^2+2f_{23}$	$.6f_{23}^2$	3.56	50	1.00	$.7f_{50}^2+2f_{50}$	$.7f_{50}^2$	1.94
24	1.00	$.7f_{24}^2+f_{24}$	$.6f_{24}^2$	1.66	51	.98	$.6f_{51}^2+7f_{51}$	$.45f_{51}^2$	0.79
25	.99	$.5f_{25}^2+.8f_{25}$	$.6f_{25}^2$	2.82	52	.99	$.9f_{52}^2+9f_{52}$	$.5f_{52}^2$	1.43
26	.99	$.6f_{26}^2+f_{26}$	$.45f_{26}^2$	3.34	53	1.00	$.55f_{53}^2+6f_{53}$	$.55f_{53}^2$	1.23
27	.99	$.7f_{27}^2+.8f_{27}$	$.4f_{27}^2$	1.24	54	.98	$.8f_{54}^2+4f_{54}$	$.5f_{54}^2$	2.28

has affected the behavior of the consumers. The demand price functions associated with the products of Firms 1 and 2 are no longer as in Scenarios 1 and 2 but are now

$$\rho_{11}(d) = -1.1d_{11} - 0.9d_{21} - 1.0d_{31} + 192,$$
$$\rho_{21}(d) = -1.2d_{21} - 0.7d_{11} - 0.8d_{31} + 176,$$
$$\rho_{31}(d) = -0.9d_{31} - 0.6d_{11} - 0.8d_{21} + 170;$$
$$\rho_{12}(d) = -0.9d_{12} - 0.8d_{22} - 0.7d_{32} + 166,$$
$$\rho_{22}(d) = -1.0d_{22} - 0.5d_{12} - 0.8d_{32} + 146,$$
$$\rho_{32}(d) = -0.8d_{32} - 0.5d_{12} - 0.6d_{22} + 153;$$
$$\rho_{13}(d) = -1.4d_{13} - 1.0d_{23} - 0.5d_{33} + 173,$$
$$\rho_{23}(d) = -1.5d_{23} - 0.4d_{13} - 0.7d_{33} + 164,$$
$$\rho_{33}(d) = -0.9d_{33} - 0.7d_{13} - 0.5d_{23} + 157.$$

The arc multipliers, the total operational and the total discarding cost functions are the same as in Scenario 2, and the new computed equilibrium link flows are reported in Table 5.3.

Table 5.3 Link multipliers, total operational cost, and total discarding cost functions and equilibrium link flow solution for Scenario 3

Link a	α_a	$\hat{c}_a(f_a)$	$\hat{z}_a(f_a)$	f_a^*	Link a	α_a	$\hat{c}_a(f_a)$	$\hat{z}_a(f_a)$	f_a^*
1	.95	$5f_1^2 + 8f_1$	$.5f_1^2$	8.42	28	.98	$.4f_{28}^2 + .8f_{28}$	$.45f_{28}^2$	0.66
2	.97	$7f_2^2 + 3f_2$	$.4f_2^2$	6.72	29	1.00	$.3f_{29}^2 + 3f_{29}$	$.55f_{29}^2$	2.29
3	.96	$6.5f_3^2 + 4f_3$	$.3f_3^2$	6.42	30	1.00	$.75f_{30}^2 + f_{30}$	$.55f_{30}^2$	1.29
4	.98	$5f_4^2 + 7f_4$	$.35f_4^2$	8.01	31	1.00	$.65f_{31}^2 + f_{31}$	$.55f_{31}^2$	1.28
5	1.00	$.7f_5^2 + f_5$	$.5f_5^2$	3.20	32	.99	$.5f_{32}^2 + 2f_{32}$	$.3f_{32}^2$	2.74
6	.99	$.9f_6^2 + 2f_6$	$.5f_6^2$	2.07	33	.99	$.4f_{33}^2 + 3f_{33}$	$.3f_{33}^2$	0.00
7	1.00	$.5f_7^2 + f_7$	$.5f_7^2$	2.73	34	1.00	$.5f_{34}^2 + 3.5f_{34}$	$.4f_{34}^2$	2.39
8	.99	$f_8^2 + 2f_8$	$.6f_8^2$	1.85	35	.98	$.4f_{35}^2 + 2f_{35}$	$.55f_{35}^2$	1.82
9	1.00	$.7f_9^2 + 3f_9$	$.6f_9^2$	2.44	36	.98	$.3f_{36}^2 + 2.5f_{36}$	$.55f_{36}^2$	0.00
10	1.00	$.6f_{10}^2 + 1.5f_{10}$	$.6f_{10}^2$	2.23	37	.99	$.55f_{37}^2 + 2f_{37}$	$.55f_{37}^2$	2.21
11	.99	$.8f_{11}^2 + 2f_{11}$	$.4f_{11}^2$	2.42	38	1.00	$.35f_{38}^2 + 2f_{38}$	$.4f_{38}^2$	3.46
12	.99	$.8f_{12}^2 + 5f_{12}$	$.4f_{12}^2$	1.75	39	1.00	$.4f_{39}^2 + 5f_{39}$	$.4f_{39}^2$	0.00
13	.98	$.9f_{13}^2 + 4f_{13}$	$.4f_{13}^2$	2.00	40	.98	$.55f_{40}^2 + 2f_{40}$	$.6f_{40}^2$	1.05
14	1.00	$.8f_{14}^2 + 2f_{14}$	$.5f_{14}^2$	2.84	41	.97	$3f_{41}^2 + 12f_{41}$	$.3f_{41}^2$	8.08
15	.99	$.9f_{15}^2 + 3f_{15}$	$.5f_{15}^2$	2.40	42	.96	$2.7f_{42}^2 + 10f_{42}$	$.4f_{42}^2$	8.13
16	1.00	$1.1f_{16}^2 + 3f_{16}$	$.6f_{16}^2$	2.60	43	.98	$1.1f_{43}^2 + 6f_{43}$	$.45f_{43}^2$	4.21
17	.98	$2f_{17}^2 + 3f_{17}$	$.45f_{17}^2$	5.02	44	.98	$.9f_{44}^2 + 5f_{44}$	$.45f_{44}^2$	3.63
18	.99	$2.5f_{18}^2 + f_{18}$	$.55f_{18}^2$	4.49	45	.97	$1.3f_{45}^2 + 6f_{45}$	$.5f_{45}^2$	4.62
19	.98	$2.4f_{19}^2 + 1.5f_{19}$	$.5f_{19}^2$	4.96	46	.99	$1.5f_{46}^2 + 7f_{46}$	$.55f_{46}^2$	3.19
20	.98	$1.8f_{20}^2 + 3f_{20}$	$.3f_{20}^2$	5.23	47	.98	$1.5f_{47}^2 + 4f_{47}$	$.4f_{47}^2$	8.60
21	.98	$2.1f_{21}^2 + 3f_{21}$	$.35f_{21}^2$	4.11	48	.98	$2.1f_{48}^2 + 6f_{48}$	$.45f_{48}^2$	6.72
22	.99	$1.9f_{22}^2 + 2.5f_{22}$	$.5f_{22}^2$	4.56	49	.99	$.6f_{49}^2 + 3f_{49}$	$.55f_{49}^2$	3.63
23	1.00	$.5f_{23}^2 + 2f_{23}$	$.6f_{23}^2$	2.44	50	1.00	$.7f_{50}^2 + 2f_{50}$	$.7f_{50}^2$	3.39
24	1.00	$.7f_{24}^2 + f_{24}$	$.6f_{24}^2$	1.47	51	.98	$.6f_{51}^2 + 7f_{51}$	$.45f_{51}^2$	1.41
25	.99	$.5f_{25}^2 + .8f_{25}$	$.6f_{25}^2$	1.02	52	.99	$.9f_{52}^2 + 9f_{52}$	$.5f_{52}^2$	1.12
26	.99	$.6f_{26}^2 + f_{26}$	$.45f_{26}^2$	2.48	53	1.00	$.55f_{53}^2 + 6f_{53}$	$.55f_{53}^2$	2.86
27	.99	$.7f_{27}^2 + .8f_{27}$	$.4f_{27}^2$	1.31	54	.98	$.8f_{54}^2 + 4f_{54}$	$.5f_{54}^2$	2.60

In Table 5.4, as we did in Chaps. 3 and 4, we report the computed path flow pattern for Scenario 3 in our case study.

Having path flow information, in addition to link flow information, is illuminating and informative for supply chain decision-makers. We can immediately see from Table 5.4 that Firm 2 supplies none of its product to demand market 2 since all of the flows on paths joining O/D pair w_2^2 are zero.

The computed equilibrium demands for the products of Firms 1, 2, and 3 are as follows:

5.4 A Case Study

Table 5.4 Computed equilibrium path flow pattern for Scenario 3

	Path definition	Path flow		Path definition	Path flow
O/D Pair $w_1^1 = (1, R_1)$	$p_1 = (1,5,17,23)$	$x_{p_1}^* = 1.87$	O/D Pair $w_1^2 = (2, R_1)$	$p_1 = (3,11,20,32)$	$x_{p_1}^* = 1.26$
	$p_2 = (1,6,18,26)$	$x_{p_2}^* = 1.46$		$p_2 = (3,12,21,35)$	$x_{p_2}^* = 0.77$
	$p_3 = (1,7,19,29)$	$x_{p_3}^* = 1.57$		$p_3 = (3,13,22,38)$	$x_{p_3}^* = 1.51$
	$p_4 = (2,8,17,23)$	$x_{p_4}^* = 0.73$		$p_4 = (4,14,20,32)$	$x_{p_4}^* = 1.63$
	$p_5 = (2,9,18,26)$	$x_{p_5}^* = 1.17$		$p_5 = (4,15,21,35)$	$x_{p_5}^* = 1.16$
	$p_6 = (2,10,19,29)$	$x_{p_6}^* = 0.87$		$p_6 = (4,16,22,38)$	$x_{p_6}^* = 2.12$
O/D Pair $w_2^1 = (1, R_2)$	$p_1 = (1,5,17,24)$	$x_{p_1}^* = 0.89$	O/D Pair $w_2^2 = (2, R_2)$	$p_1 = (3,11,20,33)$	$x_{p_1}^* = 0.00$
	$p_2 = (1,6,18,27)$	$x_{p_2}^* = 0.57$		$p_2 = (3,12,21,36)$	$x_{p_2}^* = 0.00$
	$p_3 = (1,7,19,30)$	$x_{p_3}^* = 0.66$		$p_3 = (3,13,22,39)$	$x_{p_3}^* = 0.00$
	$p_4 = (2,8,17,24)$	$x_{p_4}^* = 0.68$		$p_4 = (4,14,20,33)$	$x_{p_4}^* = 0.00$
	$p_5 = (2,9,18,27)$	$x_{p_5}^* = 0.82$		$p_5 = (4,15,21,36)$	$x_{p_5}^* = 0.00$
	$p_6 = (2,10,19,30)$	$x_{p_6}^* = 0.71$		$p_6 = (4,16,22,39)$	$x_{p_6}^* = 0.00$
O/D Pair $w_3^1 = (1, R_3)$	$p_1 = (1,5,17,25)$	$x_{p_1}^* = 0.60$	O/D Pair $w_3^2 = (2, R_3)$	$p_1 = (3,11,20,34)$	$x_{p_1}^* = 1.26$
	$p_2 = (1,6,18,28)$	$x_{p_2}^* = 0.16$		$p_2 = (3,12,21,37)$	$x_{p_2}^* = 1.05$
	$p_3 = (1,7,19,31)$	$x_{p_3}^* = 0.64$		$p_3 = (3,13,22,40)$	$x_{p_3}^* = 0.57$
	$p_4 = (2,8,17,25)$	$x_{p_4}^* = 0.49$		$p_4 = (4,14,20,34)$	$x_{p_4}^* = 1.26$
	$p_5 = (2,9,18,28)$	$x_{p_5}^* = 0.53$		$p_5 = (4,15,21,37)$	$x_{p_5}^* = 1.29$
	$p_6 = (2,10,19,31)$	$x_{p_6}^* = 0.72$		$p_6 = (4,16,22,40)$	$x_{p_6}^* = 0.54$

	Path definition	Path flow
O/D Pair $w_1^3 = (3, R_1)$	$p_1 = (41,43,47,49)$	$x_{p_1}^* = 1.87$
	$p_2 = (41,44,48,52)$	$x_{p_2}^* = 1.78$
	$p_3 = (42,45,47,49)$	$x_{p_3}^* = 0.70$
	$p_4 = (42,46,48,52)$	$x_{p_4}^* = 0.68$
O/D Pair $w_2^3 = (3, R_2)$	$p_1 = (41,43,47,50)$	$x_{p_1}^* = 1.61$
	$p_2 = (41,44,48,53)$	$x_{p_2}^* = 1.46$
	$p_3 = (42,45,47,50)$	$x_{p_3}^* = 2.07$
	$p_4 = (42,46,48,53)$	$x_{p_4}^* = 1.90$
O/D Pair $w_3^3 = (3, R_3)$	$p_1 = (41,43,47,51)$	$x_{p_1}^* = 0.84$
	$p_2 = (41,44,48,54)$	$x_{p_2}^* = 0.53$
	$p_3 = (42,45,47,51)$	$x_{p_3}^* = 1.46$
	$p_4 = (42,46,48,54)$	$x_{p_4}^* = 1.33$

$$d_{11}^* = 7.18, \quad d_{21}^* = 7.96, \quad d_{31}^* = 4.70,$$
$$d_{12}^* = 4.06, \quad d_{22}^* = 0.00, \quad d_{32}^* = 6.25,$$
$$d_{13}^* = 2.93, \quad d_{23}^* = 5.60, \quad d_{33}^* = 3.93.$$

Note that, in Scenario 3, links 33, 36, and 39 have zero flow. Hence, the *final* supply chain network topology for Scenario 3, with only those links with nonzero equilibrium flows displayed, is as depicted in Fig. 5.6. As a result of the consumers' growing inclination toward the generic substitute of the previously popular branded drug, the link flow and the demand pattern have now significantly changed. Firm 2 has lost its entire share of market 2 to its competitors, resulting in zero flows on the corresponding distribution links: 33, 36, and 39. Similarly, Firm 1 now has declining sales of its brand in demand markets 1 and 3. As noted by Johnson (2011),

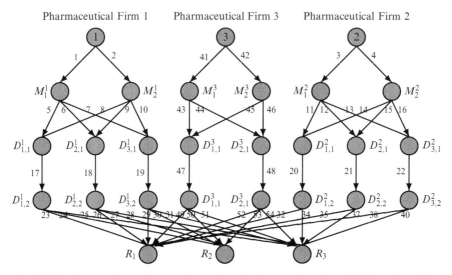

Fig. 5.6 The *final* supply chain network topology for Scenario 3 with only links with positive equilibrium flows displayed

the market share of a branded drug may decrease by as much as 40–80% after the introduction of a generic rival. Hence, our model captures the observed decrease in market share. The incurred equilibrium prices were:

$$\rho_{11}(d^*) = 172.24, \quad \rho_{21}(d^*) = 157.66, \quad \rho_{31}(d^*) = 155.09,$$
$$\rho_{12}(d^*) = 157.97, \quad \rho_{22}(d^*) = 138.97, \quad \rho_{32}(d^*) = 145.97,$$
$$\rho_{13}(d^*) = 161.33, \quad \rho_{23}(d^*) = 151.67, \quad \rho_{33}(d^*) = 148.61.$$

As expected, the introduction of the generic substitute of cholesterol regulators has also caused remarkable drops in the prices of the existing brands. Interestingly, the decrease in the price of Firm 1's product in demand markets 2 and 3 exceeds 35%. The computed profits for each of the three competitors through the production and delivery of their respective cholesterol-lowering medicines were

$$U_1(X^*) = 1,199.87, \quad U_2(X^*) = 1,062.73, \quad U_3(X^*) = 980.83.$$

Note that simultaneous declines in the amounts of demand and sales price have caused a severe reduction in the profits of Firms 1 and 2. This decline for Firm 1 is observed to be as high as 60%. The reduction in demand and prices due to the patent expiration has been observed in the sales in the market. The USA sales of Lipitor have dropped over 75% (Forbes 2012 and Fiercepharma 2012).

5.5 Outsourcing

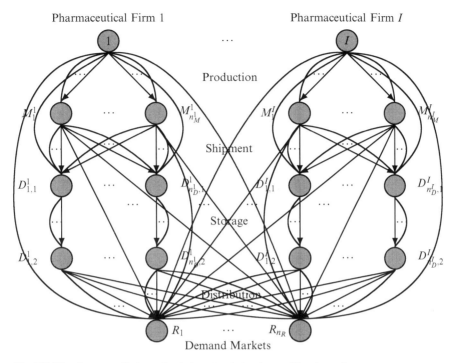

Fig. 5.7 The pharmaceutical supply chain network topology with outsourcing

5.5 Outsourcing

Not only are many of the ingredients in pharmaceutical products produced abroad but pharmaceutical firms may also outsource the production as well as other supply chain network activities. In order to model outsourcing, cf. Fig. 5.7, we expand the network in Fig. 5.1.

The additional links correspond to distinct outsourcing options, with a direct link joining a firm origin node to a destination node denoting that all the activities have been outsourced, whereas the links joining the origin (top node representing a firm) to a distribution center node representing that outsourcing the manufacturing and shipment to the distribution center is an option. The unit costs on the outsourcing links would be fixed and constant (i.e., independent of the flow), since there would be contracts involved. The model, as described in Sect. 5.2, would still be applicable, as well as the computational procedure of Sect. 5.3, but the sets of paths would now correspond to the sets of paths representing the options for each firm, as in Fig. 5.7.

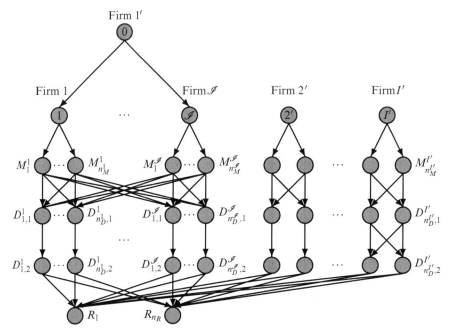

Fig. 5.8 The supply chain network post the partial merger

5.6 Mergers and Acquisitions

As mentioned earlier, pharmaceutical firms have been faced with many financial challenges. Some have used mergers and acquisitions as a strategy. Such a strategy can expand a firm's product portfolio to counter possible losses due to the expiration of patents. Zacks Equity Research (2012) suggests that the mergers and acquisitions trend will continue to grow in the pharmaceutical sector. In Fig. 5.8, we illustrate how, through an expanded network representation, a merger of several oligopolistic firms can be depicted. The underlying analytics, which can also be utilized to assess potential synergy, would be similar to that described in this chapter (cf. Nagurney 2010).

5.7 Summary and Conclusions

In this chapter, we described a supply chain network model for the study of oligopolistic competition among the producers of pharmaceuticals. The supply chain of each pharmaceutical company consisted of various activities of manufacturing, shipment, storage, and the ultimate distribution to the demand markets. The pharmaceutical firms had, as their strategies, their product flows on their supply chain networks, and hence, the competitive model is a Cournot one. The model has several novel features, which, in their totality, are significant contributions to the literature. Specifically, the contributions in this chapter are:

- An oligopolistic supply chain network model, based on variational inequality theory, that captures the perishability of pharmaceuticals through the use of arc multipliers, that assesses the discarding cost associated with the disposal of waste/perished products in the supply chain network activities, and that includes product differentiation by the consumers, capturing, for example, as to whether or not the products are branded or generic
- An adaptation of an algorithm and derivation of explicit formulae for computational purposes that yield the equilibrium product supply chain flows, the equilibrium product demands, and the incurred product prices
- A case study focused on a real-world scenario of cholesterol-lowering drugs, with the investigation of the impacts of patent expiration and the introduction of generic drug competition

We also established special cases of our model in order to reflect situations where the drugs are homogeneous or the demands for the product remain differentiated but are known and fixed, rather than elastic. In addition, we demonstrated how the model, through an expanded supply chain network topology, could also handle different outsourcing options as well as mergers and acquisitions.

5.8 Sources and Notes

This chapter is based on the paper by Masoumi et al. (2012). In this chapter, we have synthesized the relationships of this model with other perishable product supply chain network models discussed in this book. We also brought the case study up-to-date and presented additional output data and discussions. In addition, we developed an extension of the model to include outsourcing through an expanded supply chain network topology and mergers and acquisitions.

It is also worth mentioning that in perishable product (as well as other product) supply chains, resources, such as distribution centers, may have to be shared. The supply chain network topology depicting such an economic reality is depicted in Fig. 5.9.

We now highlight some of the related literature. Papageorgiou et al. (2001), Gatica et al. (2003), Amaro and Barbosa-Povoa (2008), Tsiakis and Papageorgiou (2008), and Sousa et al. (2008) applied mixed-integer linear programming techniques to solve various problems of planning, capacity allocation, and distribution of medication drugs. Papageorgiou (2009) and Yu et al. (2010) surveyed the challenges and methodologies in the area of pharmaceutical supply chains. Subramanian et al. (2001) developed an integrated optimization-simulation framework to resolve the uncertainties in the pipeline management problem. Niziolek (2008), in her thesis, applied simulation techniques to study various shipment strategies in medical drug

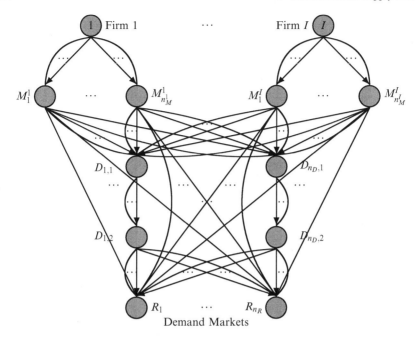

Fig. 5.9 The supply chain network topology of the oligopoly with distribution center sharing

supply networks. Recently, newly applied technologies in the area of operations of pharmaceutical chains, including RFID-based frameworks, have been studied in the literature (see Yue et al. 2008 and Schapranow et al. 2011). Rossetti et al. (2011) described the complexities of pharmaceutical supply chains and provided insights into this industry and its challenges.

References

Amaro ACS, Barbosa-Povoa APFD (2008) Planning and scheduling of industrial supply chains with reverse flows: A real pharmaceutical case study. Comput Chem Eng 32(11):2606–2625

Business Wire (2009) CVS has consumers going nuts over expired milk, eggs, infant formula, medicine, according to Change to Win. April 2. Available online at: http://www.businesswire.com/news/google/20090402006264/en

Dafermos S, Nagurney A (1987) Oligopolistic and competitive behavior of spatially separated markets. Reg Sci Urban Econ 17:245–254

De la Garza D (2011) Pharmaceutical patent protection and why 2012 is the year companies fear the most. The Strategic Sourceror, September 1.

Available online at: http://www.strategicsourceror.com/2011/09/pharmaceutical-patent-protection-and.html

Dunehew A (2005) Changing dynamics in the pharmaceutical supply chains: A GPO perspective. Am J Health Syst Pharm 62(5):527–529

Emanuel EJ (2011) Shortchanging cancer patients. The New York Times, August 6. Available online at: http://www.nytimes.com/2011/08/07/opinion/sunday/ezekiel-emanuel-cancer-patients.html

Fiercepharma (2012) Lipitor copy powers a surge in Ranbaxy sales. May 9. Available online at: http://fiercepharma.com/story/lipitor-copy-powers-surge-ranbaxy-sales-profits/2012-05-09

Forbes (2012) Pfizer Q2 profit up 25% despite stinging loss of Lipitor July 21. Available online at: http://www.forbes.com/sites/abrambrown/2012/07/31/pfizer-q2-profit-up-25-despite-stinging-loss-of-lipitor/

Gatica G, Papageorgiou LG, Shah N (2003) Capacity planning under uncertainty for the pharmaceutical industry. Chem Eng Res Des 81(6):665–678

Harris G (2011) Deal in place for inspecting foreign drugs. The New York Times, August 13. Available online at: http://www.nytimes.com/2011/08/13/science/13drug.html?scp=1&sq=deal%20in%20place%20for%20inspecting%20foreign%20drugs&st=cse

Johnson T (2011) The debate over generic-drug trade. Council on Foreign Relations, August 3. Available online at: http://www.cfr.org/drugs/debate-over-generic-drug-trade/p18055#p2

Karaesmen IZ, Scheller-Wolf A, Deniz B (2011) Managing perishable and aging inventories: Review and future research directions. In: Kempf K, Keskinocak P, Uzsoy P (eds) Planning production and inventories in the extended enterprise. Springer, Berlin, pp 393–436

Marucheck A, Greis N, Mena C, Cai L (2011) Product safety and security in the global supply chain: Issues, challenges and research opportunities. J Oper Manag 29(7–8):707–720

Masoumi AH, Yu M, Nagurney A (2012) A supply chain generalized network oligopoly model for pharmaceuticals under brand differentiation and perishability. Transport Res E 48(4):762–780

Mihalopoulos D (2009) Expired or lost drugs costing city $1 million. Chicago Tribune, February 3

Muller J, Popke C, Urbat M, Zeier A, Plattner H (2009) A simulation of the pharmaceutical supply chain to provide realistic test data. First International Conference on Advances in System Simulation, Porto, Portugal

Nagurney A (2010) Formulation and analysis of horizontal mergers among oligopolistic firms with insights into the merger paradox: A supply chain network perspective. Comput Manag Sci 7(4):377–406

Nagurney A, Li D (2012) A dynamic network oligopoly model with transportation costs, product differentiation, and quality competition. Isenberg School of Management, University of Massachusetts, Amherst, MA

Nagurney A, Yu M (2012) A supply chain network game theoretic framework for time-based competition with transportation costs and product differentiation. Isenberg School of Management, University of Massachusetts, Amherst, MA

Niziolek L (2008) A Simulation-based Study of Distribution Strategies for Pharmaceutical Supply Chains. PhD Thesis, Purdue University, West Lafayette, Indiana

Papageorgiou LG (2009) Supply chain optimisation for the process industries: Advances and opportunities. Comput Chem Eng 33(12):1931–1938

Papageorgiou LG, Rotstein GE, Shah N (2001) Strategic supply chain optimization for the pharmaceutical industries. Ind Eng Chem Res 40(1):275–286

Rossetti CL, Handfield R, Dooley KJ (2011) Forces, trends, and decisions in pharmaceutical supply chain management. Int J Phys Distrib Logist Manag 41(6):601–622

Rossi K (2011) Several prescription drug patents set to expire. September 13. Available online at: http://www.krtv.com/news/several-prescription-drug-patents-set-to-expire

Schapranow M-P, Zeier A, Plattner H (2011) A formal model for enabling RFID in pharmaceutical supply chains. Proceedings of the 44th Hawaii International Conference on System Sciences. doi:10.1109/HICSS.2011.10

Schneider ME (2011) Profiteers take advantage of drug shortages. Family Practice News, August 16. Available online at: http://www.familypracticenews.com/news/more-top-news/single-view/profiteers-take-advantage-of-drug-shortages/c5d7e6858b.html

Shah N (2004) Pharmaceutical supply chains: Key issues and strategies for optimisation. Comput Chem Eng 28(6–7):929–941

Sousa R, Shah N, Papageorgiou LG (2008) Supply chain design and multilevel planning: An industrial case. Comput Chem Eng 32(11):2643–2663

Subramanian D, Pekny JF, Reklaitis GV (2001) A simulation-optimization framework for research and development pipeline management. Am Inst Chem Eng J 47(10):2226–2242

Szabo L (2011) Drug shortages set to reach record levels. USA Today, August 15. Available online at: http://yourlife.usatoday.com/health/story/2011/08/Drug-shortages-set-to-reach-record-levels/49984446/1

Tsiakis P, Papageorgiou LG (2008) Optimal production allocation and distribution supply chain networks. Int J Prod Econ 111(2):468–483

WPRI (2009) CVS offers $2 for expired products. June 11. Available online at: http://www.wpri.com/dpp/news/local_news/local_wpri_money_for_expired_products_in_ca_pharmacies_20090611

Yu X, Li C, Shi Y, Yu M (2010) Pharmaceutical supply chain in China: Current issues and implications for health system reform. Health Pol 97(1):8–15

Yue D, Wu X, Bai J (2008) RFID application framework for pharmaceutical supply chain. Proceedings of IEEE International Conference on Service Operations and Logistics, and Informatics, 1125–1130, Beijing, China

Zacks Equity Research (2012) Pharma & biotech stock outlook –April 2012

Chapter 6
Fast Fashion Apparel Supply Chains

Abstract In this chapter, we develop an oligopoly model for fashion supply chains in the case of differentiated products with the inclusion of environmental concerns and with supply chain network link time consumption functions. Each fashion firm seeks to maximize its profit, with generalized cost components, as well as to minimize its emissions throughout its supply chain with the latter criterion being weighted in an individual manner by each firm. A case study illustrates both the generality of the modeling framework as well as how the computational scheme can be used in practice to explore the effects of changes in the demand functions; in the total generalized cost and the total emission functions, as well as in the weights.

6.1 Motivation and Overview

In earlier chapters, we presented perishable product supply chain network models, associated algorithms, and applications ranging from blood banking systems to medical nuclear supply chains to food supply chains and pharmaceutical supply chains. In this final chapter, we explore the world of fast fashion apparel supply chains through our integrated methodology.

During the past two decades, fashion retailers, such as Benetton, H&M, Topshop, and Zara, have revolutionized the fashion industry by following what has become known as the *fast fashion* strategy, in which retailers respond to shifts in the market within just a few weeks versus an industry average of 6 months. Specifically, fast fashion is a concept developed in Europe to serve markets for teenage and young adult women who desire trendy, short-cycle, and relatively inexpensive clothing. Fast fashion chains have grown quicker than the industry as a whole and have seized market share from traditional rivals, since they aim to obtain fabrics, to manufacture samples and to start shipping products with far shorter lead times than those of the traditional production calendar (see Sull and Turconi 2008 and Doeringer and Crean 2006).

Nordas et al. (2006) argued that time is a critical component in the case of labor-intensive products such as clothing as well as consumer electronics, both examples of classes of products that are increasingly time-sensitive. They presented two case studies of the textile and clothing sector in Bulgaria and the Dominican Republic, respectively. They noted that, despite higher production costs than in China, their closeness to major markets gave these two countries the advantage of a shorter lead time that allowed them to specialize in fast fashion products. Interestingly, they also identified that lengthy, time-consuming administrative procedures for exports and imports reduce the probability that firms will even enter export markets for time-sensitive products. Clearly, superior time performance must be weighed against the associated costs. Indeed, as noted by So (2000), it can be costly to deliver superior time performance, since delivery time performance generally depends on the available capacity and on the operating efficiency of the system. It is increasingly evident that, in the case of time-sensitive products, with fashion being an example par excellence, an appropriate supply chain network analytics framework for such products must capture both the operational (and other) cost dimension as well as the time dimension.

In addition to time pressures, the fashion and apparel industry also faces immense challenges as well as opportunities in the reduction of its environmental impact globally. The two issues of time and sustainability are actually intimately related since faster deliveries may involve more energy-intensive transportation and associated higher emissions. Moreover, consumers are becoming increasingly aware of such issues through their demand for apparel that is produced and distributed in a manner that minimizes the use (and discarding) of toxic dyes, raw materials such as cotton grown with pesticides, as well as the generation of waste in terms of textiles and byproducts (including packaging). Hence, firms such as Levi's, Gap, H&M, and Wal-Mart that wish to enhance or to maintain a positive brand identity are responding accordingly and becoming increasingly environmentally conscious (see, e.g., Claudio 2007, Glausiusz 2008, Rosenbloom 2010, and Tucker 2010).

In order to fix ideas, and to emphasize the scope of the environmental issues associated with the fashion and apparel industry, we now provide some details on the fashion industry. Polyester is a man-made fiber whose demand from the fashion industry has doubled in the past 15 years. Its manufacture requires petroleum and releases such emissions as volatile organic compounds and gases such as hydrogen chloride, as well as particulates. Additional byproducts associated with its production are emitted in the waste water. However, even natural fibers used in textiles for apparel may also leave a large environmental imprint. For example, the production of cotton, one of the most versatile fibers used in clothing, accounts for a quarter of all the pesticides used in the United States, which is the largest exporter of cotton in the world (see Claudio 2007).

According to the National Resources Defense Counsil (NRDC) (see Tucker 2010), textile manufacturing pollutes as much as 200 tons of water per ton of fabric. In China, for example, a textile factory may burn about 7 tons of carbon emitting coal per ton of fabric produced. In the case of blue jean production, Xintang, located in the northeastern part of the Pearl River Delta in China, is where approximately

6.1 Motivation and Overview

200 million pairs of jeans are produced annually for 1,000 different labels. The standard jean dyeing process dispenses into its waste water a mixture of dye, bleach, and detergent. As a consequence, the production of blue jeans in such a manner is partly to blame for the pollution of the Pearl River (see UPI.com 2010).

As the production of apparel has become global and competition has intensified, with an increased prominence of brands and buyer-driven value chains, new networks are transforming this industry. Whereas in 1992 about 49% of all retail apparel sold in the USA was actually made there, by 1999 the proportion had fallen to just 12% (Rabon 2001). Between 1990 and 2000, the value of apparel imports to the USA increased from $25 billion to $64 billion. According to Gereffi and Memedovic (2003), the top exporters of apparel in 2000, with a value of over $1 billion in US dollars, were China, Hong Kong, the United States, Mexico, and Turkey, whereas in 1980, the major exporters were Hong Kong, South Korea, Taiwan, China, and the United States. However, as noted in Nagurney and Woolley (2010), with the growing investment and industrialization in developing nations, it is also important to evaluate the overall impact at not only the operational level but also in terms of the environment. For example, between 1988 and 1995, multinational corporations invested nearly $422 billion worth in new factories, supplies, and equipment in developing countries (World Resources Institute 1998). Through globalization, firms of industrialized nations may make use of manufacturing plants in developing nations that offer lower production costs; however, more than not, combined with inferior environmental concerns, due to a looser environmental regulatory system and/or lower environmental impact awareness.

It is imperative to account for the environmental emissions associated with the fashion and apparel industry, especially given its global dimensions in terms of both manufacturing plant locations and demand markets. The accounting should include the emissions generated in the transportation and distribution of the products across oceans and continents. For example, H&M is cognizant of the environmental impact of even the fuels used in the transportation of its fashion products as well as the number of shipments needed for distribution. According to the Guardian (2010), H&M identified that 51% of its carbon imprint in 2009 was due to transportation. To reduce the associated emissions, it began more direct shipments that avoided intermediate warehouses, decreased the volumes shipped by ocean and air by 40%, and increased the volume of products shipped by rail, resulting in an over 700 ton decrease in the amount of carbon dioxide emitted.

In this chapter, we present an oligopoly model for fashion supply chain competition which explicitly considers different brands and different degrees of environmental consciousness and sustainability, along with activity time consumption. As in the models in Chaps. 4 and 5, the network-based model is formulated and studied as a variational inequality problem and captures competition among the firms in manufacturing, transportation/distribution, and storage. However, the model in this chapter is no longer a generalized network model, but, rather, a pure network model, since all the arc and, hence, the path and the arc–path multipliers are equal to one. In the model in this chapter, time is captured through explicit link time consumption functions and with the use of link total generalized costs.

In this chapter, we assume that the firms seek not only to maximize their profits but also care, in an individual way, about the emissions that they generate. The supply chain network oligopoly model that we present has the following features:

- It handles product differentiation through branding.
- It explicitly allows for alternative modes of transportation for product distribution as well as the possibility of an option of direct shipment from manufacturing plants.
- It incorporates supply chain network activity time consumption functions.
- It enables each fashion/apparel producing firm to individually determine, by use of its individual concern through a weighting factor, its environmental impacts through the emissions that it generates not only in the manufacture of its product but throughout its supply chain, with the ultimate deliveries at the demand markets.

This chapter is organized as follows. In Sect. 6.2, we describe the sustainable fashion supply chain network oligopoly model with brand differentiation, relate it to other models in this book, and provide some qualitative properties. In Sect. 6.3, we present the computational procedure which we then apply in Sect. 6.4 to compute solutions to a spectrum of numerical examples that comprise our case study. The case study illustrates both the generality of our framework and its applicability. In Sect. 6.5, we summarize our findings and present our conclusions. Section 6.6 contains the Sources and Notes.

6.2 The Sustainable Fashion Supply Chain Network Oligopoly Model

We consider I fashion firms, with a typical firm denoted by i, who are involved in the production, storage, and distribution of a fashion product and who compete noncooperatively in an oligopolistic manner. As in the game theoretic model in Chap. 5, each firm corresponds to an individual brand representing the product that it produces.

Each fashion firm is represented as a network of its economic activities (cf. Fig. 6.1). Note that the supply chain network in Fig. 6.1 has a topology identical to the pharmaceutical supply chain network in Fig. 5.1, except for one distinguishing feature. In the network in Fig. 6.1, there are no multiple links associated with the storage links. Apparel, unlike pharmaceutical products, does not require special storage conditions.

Each fashion firm seeks to determine its optimal product quantities by using Fig. 6.1 as a schematic. Each fashion firm i, hence, is considering n_M^i manufacturing facilities/plants, n_D^i distribution centers, and serves the same n_R demand markets. Let L^i denote the set of directed links representing the economic activities associated with firm i. As in the preceding modeling chapters, let $\mathscr{G} = [N, L]$ denote the graph consisting of the set of nodes N and the set of links L in Fig. 6.1, where $L \equiv \cup_{i=1,\ldots,I} L^i$.

6.2 The Sustainable Fashion Supply Chain Network Oligopoly Model

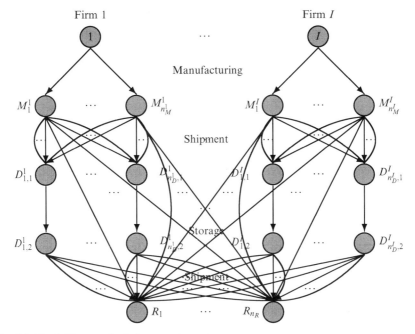

Fig. 6.1 The fashion supply chain network topology

The links from the top-tiered nodes i; $i = 1,\ldots,I$, in Fig. 6.1 are connected to the manufacturing nodes of the respective firm i, which are denoted, respectively, by $M_1^i,\ldots,M_{n_M^i}^i$. These links represent the manufacturing links. The links from the manufacturing nodes, in turn, are connected to the distribution center nodes of each firm i, which are denoted by $D_{1,1}^i,\ldots,D_{n_D^i,1}^i$. These links correspond to the shipment links between the manufacturing plants and the distribution centers where the product is stored. Observe that there are alternative shipment links to denote different possible modes of transportation (which would also have associated with them different levels of emissions). Alternative modes of transportation may include rail, air, truck, sea, as appropriate.

The links joining nodes $D_{1,1}^i,\ldots,D_{n_D^i,1}^i$ with nodes $D_{1,2}^i,\ldots,D_{n_D^i,2}^i$ correspond to the storage links. Finally, there are alternative shipment links joining the nodes $D_{1,2}^i,\ldots,D_{n_D^i,2}^i$ with the demand market nodes: $R_1,\ldots,R_{n_R}$. Here we also allow for multiple modes of transportation, as depicted using multiple arcs in Fig. 6.1.

In addition, in order to represent another option, as was noted for H&M in Sect. 6.1, we allow for the possibility that a firm may wish to have the product transported directly from a manufacturing plant to a demand market, using one or more transportation shipment modes.

We emphasize that the network topology in Fig. 6.1 is only representative, for definiteness. In fact, the model can handle any prospective supply chain network topology provided that there is a top-tiered node to represent each firm and bottom-tiered nodes to represent the demand markets with a sequence of directed links. We assume that there is at least one path joining each top-tiered node with each bottom-tiered node. Hence, different supply chain network topologies to that depicted in Fig. 6.1 correspond to distinct fashion supply chain network problems.

Let d_{ik} denote the demand for fashion firm i's product at demand market R_k. Note that, in our model, we consider the general situation where the products of all the fashion firms are not homogeneous but are differentiated by *brand*.

The notation is similar to that used in Chaps. 2 through 5. Let x_p denote the nonnegative flow on path p joining (origin) node i; $i = 1, \ldots, I$ with a (destination) demand market node. Then the following conservation of flow equations must hold:

$$\sum_{p \in \mathscr{P}_k^i} x_p = d_{ik}, \quad i = 1, \ldots, I; k = 1, \ldots, n_R, \tag{6.1}$$

where $\mathscr{P}_k^i$ denotes the set of all paths joining the origin node i with destination node R_k, and $\mathscr{P} \equiv \cup_{i=1,\ldots,I} \cup_{k=1,\ldots,n_R} \mathscr{P}_k^i$, denotes the set of all paths in Fig. 6.1. According to (6.1), the demand for fashion firm i's product at demand point R_k must be equal to the sum of the product flows from firm i to that demand market. We group the demands d_{ik}; $i = 1, \ldots, I$; $k = 1, \ldots, n_R$ into the $I \times n_R$-dimensional vector d. Note that, in the model in this chapter, we no longer make use of the arc and path multipliers, as we did in Chaps. 2 through 5. This is because, although fashion apparel is time-sensitive, there are no losses in the chain. Hence, all the arc multipliers α_a; $a \in L$ are equal to 1, as are the path multipliers μ_p, $p \in \mathscr{P}$. The arc–path multipliers α_{ap} then become δ_{ap}, where $\delta_{ap} = 1$, if link a is contained in path p, and $\delta_{ap} = 0$, otherwise.

For the sake of generality, we refer to the bottom-tiered nodes in Fig. 6.1 as demand markets. Of course, they may correspond to retailers.

We assume that there is a demand price function (sometimes also referred to as the inverse demand function) associated with each fashion firm's product at each demand market. We denote the demand price of fashion firm i's product at demand market R_k by ρ_{ik} and we assume, as given, the demand price functions:

$$\rho_{ik} = \rho_{ik}(d), \quad i = 1, \ldots, I; k = 1, \ldots, n_R. \tag{6.2}$$

Hence, the price for fashion firm i's product at a particular demand market may depend not only on the demands for this fashion product at that and the other demand markets but also on the demands for the other substitutable fashion products at all the demand points. The functions in (6.2) capture competition on the demand side of the competitive fashion supply chain network. Such demand price functions are of the form that we have utilized in Chaps. 4 and 5 and are used in differentiated oligopolies (cf. Shy 1996 and Carlton and Perloff 2004). However, these demand price functions are not limited to being linear, as is commonly assumed in economics. We assume that the demand price functions are continuous, continuously differentiable, and monotone decreasing.

6.2 The Sustainable Fashion Supply Chain Network Oligopoly Model

Let f_a denote the flow on link a. We must have the following conservation of flow equations satisfied:

$$f_a = \sum_{p \in \mathcal{P}} x_p \delta_{ap}, \quad \forall a \in L, \quad (6.3)$$

where $\delta_{ap} = 1$ if link a is contained in path p and $\delta_{ap} = 0$, otherwise. In other words, the flow on a link is equal to the sum of flows on paths that contain that link. Unlike the models in the preceding chapters of this book, the volume of product that flows from a node in the supply chain network on a link is the amount that arrives at its successor node.

Hence, the model in this chapter is a pure network model since all the arc and path multipliers are equal to one and, therefore, they do not appear in the conservation of flow equations (6.1) and (6.3).

The path flows must be nonnegative, that is,

$$x_p \geq 0, \quad \forall p \in \mathcal{P}. \quad (6.4)$$

As we did in Chaps. 2 through 5, we group the path flows into the vector $x \in R_+^{n_\mathcal{P}}$.

The total operational cost on a link, be it a manufacturing/production link, a shipment link, or a storage link may, in general, be a function of the product flows on all the links, that is,

$$\hat{c}_a = \hat{c}_a(f), \quad \forall a \in L, \quad (6.5)$$

where f is the vector of all the link flows. The above total cost expressions capture competition among the firms for resources used in manufacturing, transportation, and storage of their fashion products. We assume that the total operational cost on each link is convex and is continuously differentiable.

We also assume that, in addition to a total operational cost function on each link, there is also a total time consumption function $\hat{t}_a$, $\forall a \in L$, such that

$$\hat{t}_a = \hat{t}_a(f), \quad \forall a \in L, \quad (6.6)$$

with similar properties to those of the link total operational cost functions in (6.5).

Each firm i determines how it weights time, with a nonnegative weight ω_i^1 associated with its link total time functions, yielding a *generalized* link cost function $\hat{g}_a$ constructed for each link $a \in L^i$ as

$$\hat{g}_a(f) = \hat{c}_a(f) + \omega_i^1 \hat{t}_a(f), \quad \forall a \in L^i. \quad (6.7)$$

A generalized cost captures not only the operational cost but, here, also includes a monetarized value of time (see Nagurney and Yu 2011). Hence, unlike in the competitive game theory models in Chaps. 4 and 5, here we capture time (and perishability) not through arc multipliers but through the objective functions of the firms, as described below.

As in Chaps. 4 and 5, we let X_i denote the vector of strategy variables associated with firm i where X_i is the vector of path flows associated with firm i, that is,

$X_i \equiv \{\{x_p\} | p \in \mathscr{P}^i\} \in R_+^{n_{\mathscr{P}^i}}$, where $\mathscr{P}^i \equiv \cup_{k=1,\ldots,n_R} \mathscr{P}_k^i$, and $n_{\mathscr{P}^i}$ denotes the number of paths from firm i to the demand markets. X is then the vector of all the firms' strategies, that is, $X \equiv \{\{X_i\} | i = 1, \ldots, I\}$.

The profit function π_i of firm i is the difference between the firm's revenue and its total generalized link costs, that is,

$$\pi_i = \sum_{k=1}^{n_R} \rho_{ik}(d) \sum_{p \in \mathscr{P}_k^i} x_p - \sum_{a \in L^i} \hat{g}_a(f). \tag{6.8}$$

All the fashion firms are concerned with their environmental impacts along their supply chains, but, possibly, to different degrees. The emission–generation function associated with link a, denoted by $\hat{e}_a$, is assumed to be a function of the product flow on that link, that is,

$$\hat{e}_a = \hat{e}_a(f_a), \quad \forall a \in L. \tag{6.9}$$

These functions are assumed to be convex and continuously differentiable.

Each fashion firm aims to minimize the total amount of emissions generated in the manufacture, storage, and shipment of its product. Hence, the other objective of firm i is given by

$$\text{Minimize} \sum_{a \in L^i} \hat{e}_a(f_a). \tag{6.10}$$

We can now construct another weighted function, which we refer to as the utility function, often referred to as a *value* function, associated with the two criteria faced by each firm. The term ω_i^2 is assumed to be the price that firm i would be willing to pay for each unit of emission on each of its links. This term, hence, represents the environmental concern of firm i, with a higher ω_i^2 denoting a greater concern for the environment. Consequently, the multicriteria decision-making problem faced by fashion firm i is to

$$\text{Maximize} \quad U_i = \sum_{k=1}^{n_R} \rho_{ik}(d) \sum_{p \in \mathscr{P}_k^i} x_p - \sum_{a \in L^i} \hat{g}_a(f) - \omega_i^2 \sum_{a \in L^i} \hat{e}_a(f_a) \tag{6.11}$$

subject to the constraints: (6.1), (6.3), and (6.4).

Note that, in view (6.1)–(6.11), we may express each U_i solely in terms of path flows by substituting the respective expressions for the demands in terms of the path flows (cf. (6.1)) and the link flows in terms of the path flows (cf. (6.3)) into each U_i; $i = 1, \ldots, I$ given by (6.11). This means that only the path flow constraints (6.4) remain. Hence, we may write

$$U = U(X), \tag{6.12}$$

where U is the I-dimensional vector of all the firms' utilities.

As in Chaps. 4 and 5, we consider the usual oligopolistic market mechanism in which the I firms select their product path flows (which correspond to quantity decision variables in the Cournot oligopoly framework) in a noncooperative manner, each one trying to maximize its own utility. We seek to determine a path flow pattern

X^* for which the I firms will be in a state of equilibrium as defined in Definition 4.1 in Chap. 4. Note that, according to Definition 4.1, an equilibrium is established if no firm can individually improve its utility by changing its production path flows given the production path flow decisions of the other firms.

The variational inequality formulations of the Cournot–Nash sustainable fashion supply chain network problem satisfying Definition 4.1, in both path flows and link flows, respectively, are given in the following theorem.

Theorem 6.1 (Variational Inequality Formulations). *Assume that for each fashion firm i; $i = 1, \ldots, I$, the utility function $U_i(X)$ is concave with respect to the variables in X_i and is continuously differentiable. Then $X^* \in K$ is a sustainable fashion supply chain network Cournot–Nash equilibrium according to Definition 4.1 if and only if it satisfies the variational inequality:*

$$-\sum_{i=1}^{I} \langle \nabla_{X_i} U_i(X^*), X_i - X_i^* \rangle \geq 0, \quad \forall X \in K, \tag{6.13}$$

where $\langle \cdot, \cdot \rangle$ denotes, as before, the inner product in the corresponding Euclidean space and $\nabla_{X_i} U_i(X)$ denotes the gradient of $U_i(X)$ with respect to X_i. The solution of variational inequality (6.13) is equivalent to the solution of the variational inequality: determine $x^* \in K^1$ satisfying

$$\sum_{i=1}^{I} \sum_{k=1}^{n_R} \sum_{p \in \mathscr{P}_k^i} \left[\frac{\partial \hat{G}_p(x^*)}{\partial x_p} + \omega_i^2 \frac{\partial \hat{E}_p(x^*)}{\partial x_p} - \rho_{ik}(x^*) - \sum_{l=1}^{n_R} \frac{\partial \rho_{il}(x^*)}{\partial d_{ik}} \sum_{q \in \mathscr{P}_l^i} x_q^* \right]$$
$$\times [x_p - x_p^*] \geq 0, \quad \forall x \in K^1, \tag{6.14}$$

where $K^1 \equiv \{x | x \in R_+^{n_{\mathscr{P}}}\}$, and for each path p; $p \in \mathscr{P}_k^i$; $i = 1, \ldots, I$; $k = 1, \ldots, n_R$: $\frac{\partial \hat{G}_p(x)}{\partial x_p} \equiv \sum_{b \in L^i} \sum_{a \in L^i} \frac{\partial \hat{g}_b(f)}{\partial f_a} \delta_{ap}$, $\frac{\partial \hat{E}_p(x)}{\partial x_p} \equiv \sum_{a \in L^i} \frac{\partial \hat{e}_a(f_a)}{\partial f_a} \delta_{ap}$.

Variational Inequality (6.14) can be reexpressed in terms of link flows as: determine the vector of equilibrium link flows and the vector of equilibrium demands $(f^*, d^*) \in K^2$, such that

$$\sum_{i=1}^{I} \sum_{a \in L^i} \left[\sum_{b \in L^i} \frac{\partial \hat{g}_b(f^*)}{\partial f_a} + \omega_i^2 \frac{\partial \hat{e}_a(f_a^*)}{\partial f_a} \right] \times [f_a - f_a^*]$$
$$+ \sum_{i=1}^{I} \sum_{k=1}^{n_R} \left[-\rho_{ik}(d^*) - \sum_{l=1}^{n_R} \frac{\partial \rho_{il}(d^*)}{\partial d_{ik}} d_{il}^* \right] \times [d_{ik} - d_{ik}^*] \geq 0, \quad \forall (f,d) \in K^2, \tag{6.15}$$

where $K^2 \equiv \{(f,d) | \exists x \geq 0, \text{ and } (6.1) \text{ and } (6.3) \text{ hold}\}$.

Proof. Variational inequality (6.13) follows from the first part of the proof of Theorem 4.1. Observe now that

$$\nabla_{X_i} U_i(X) = \left[\frac{\partial U_i}{\partial x_p}; p \in \mathscr{P}_k^i; k = 1, \ldots, n_R \right], \tag{6.16}$$

where for each path p; $p \in \mathscr{P}_k^i$:

$$\frac{\partial U_i}{\partial x_p} = \frac{\partial \left[\sum_{l=1}^{n_R} \rho_{il}(d) \sum_{q \in \mathscr{P}_l^i} x_q - \sum_{b \in L^i} \hat{g}_b(f) - \omega_i^2 \sum_{b \in L^i} \hat{e}_b(f_b)\right]}{\partial x_p}$$

$$= \sum_{l=1}^{n_R} \frac{\partial \left[\rho_{il}(d) \sum_{q \in \mathscr{P}_l^i} x_q\right]}{\partial x_p} - \frac{\partial \left[\sum_{b \in L^i} \hat{g}_b(f)\right]}{\partial x_p} - \omega_i^2 \frac{\partial \left[\sum_{b \in L^i} \hat{e}_b(f_b)\right]}{\partial x_p}$$

$$= \rho_{ik}(d) + \sum_{l=1}^{n_R} \frac{\partial \rho_{il}(d)}{\partial d_{ik}} \frac{\partial d_{ik}}{\partial x_p} \sum_{q \in \mathscr{P}_l^i} x_q - \sum_{a \in L^i} \frac{\partial \left[\sum_{b \in L^i} \hat{g}_b(f)\right]}{\partial f_a} \frac{\partial f_a}{\partial x_p}$$

$$- \omega_i^2 \sum_{a \in L^i} \frac{\partial \left[\sum_{b \in L^i} \hat{e}_b(f_b)\right]}{\partial f_a} \frac{\partial f_a}{\partial x_p}$$

$$= \rho_{ik}(d) + \sum_{l=1}^{n_R} \frac{\partial \rho_{il}(d)}{\partial d_{ik}} \sum_{q \in \mathscr{P}_l^i} x_q - \sum_{a \in L^i} \sum_{b \in L^i} \frac{\partial \hat{g}_b(f)}{\partial f_a} \delta_{ap}$$

$$- \omega_i^2 \sum_{a \in L^i} \frac{\partial \hat{e}_a(f_a)}{\partial f_a} \delta_{ap}. \tag{6.17}$$

The demand price functions (6.2) can be rewritten in light of (6.1) as functions of path flows. By making use then of the definitions of $\frac{\partial \hat{G}_p(x)}{\partial x_p}$ and $\frac{\partial \hat{E}_p(x)}{\partial x_p}$ above, variational inequality (6.14) is immediate. In addition, the equivalence between variational inequalities (6.14) and (6.15) can be proved with (6.1) and (6.3). □

Variational inequalities (6.14) and (6.15) can be put into standard form (2.41). The vector X has already been defined as the vector of path flows. We now define the vector

$$F(X) \equiv \left[\frac{\partial \hat{G}_p(x)}{\partial x_p} + \omega_i^2 \frac{\partial \hat{E}_p(x)}{\partial x_p} - \rho_{ik}(x) - \sum_{l=1}^{n_R} \frac{\partial \rho_{il}(x)}{\partial d_{ik}} \sum_{q \in \mathscr{P}_l^i} x_q ; \right.$$
$$\left. p \in \mathscr{P}_k^i; i = 1, \ldots, I; k = 1, \ldots, n_R \right], \tag{6.18}$$

and $\mathscr{K} \equiv K^1$. Then (6.14) can be rewritten as (2.41). If we define the vectors $X \equiv (f,d)$ and $F(X) \equiv (F_1(X), F_2(X))$, such that

$$F_1(X) = \left[\sum_{b \in L^i} \frac{\partial \hat{g}_b(f)}{\partial f_a} + \omega_i^2 \frac{\partial \hat{e}_a(f_a)}{\partial f_a}; a \in L^i; i = 1, \ldots, I\right],$$

$$F_2(X) = \left[-\rho_{ik}(d) - \sum_{l=1}^{n_R} \frac{\partial \rho_{il}(d)}{\partial d_{ik}} d_{il}; i = 1, \ldots, I; k = 1, \ldots, n_R\right], \tag{6.19}$$

and $\mathscr{K} \equiv K^2$, then (6.15) can be reexpressed as (2.41).

Qualitative properties of existence and uniqueness can be obtained using similar assumptions and arguments as in Theorems 4.2 and 4.3 in Chap. 4.

6.3 The Algorithm

The Euler method, which we have utilized for computations in Chaps. 2, 4, and 5, will also be used here for the solution of the fast fashion apparel supply chain network problems in our case study in Sect. 6.4 because of the nice structure of the induced problems at each iteration. Please refer to Chap. 2 for conditions for convergence.

6.3.1 Explicit Formulae for the Euler Method Applied to the Sustainable Fashion Supply Chain Network Oligopoly Variational Inequality (6.14)

The elegance of this procedure for the computation of solutions to the sustainable fashion supply chain network oligopoly problem modeled in Sect. 6.2 can be seen in the following explicit formulae. In particular, (2.59) for the sustainable fashion supply chain network oligopoly model governed by variational inequality problem (6.14) yields the following closed form expression for the fashion product path flow on each path $p \in \mathscr{P}_k^i; i = 1, \ldots, I; k = 1, \ldots, n_R$, at iteration $\tau + 1$:

$$x_p^{\tau+1} = \max\left\{0, x_p^\tau + a_\tau \left(\rho_{ik}(x^\tau) + \sum_{l=1}^{n_R} \frac{\partial \rho_{il}(x^\tau)}{\partial d_{ik}} \sum_{q \in \mathscr{P}_l^i} x_q^\tau - \frac{\partial \hat{G}_p(x^\tau)}{\partial x_p} - \omega_i^2 \frac{\partial \hat{E}_p(x^\tau)}{\partial x_p}\right)\right\}. \tag{6.20}$$

In the next section, we solve sustainable fashion supply chain network oligopoly problems using the above algorithmic scheme.

6.4 A Case Study

In this section, we present a case study in which we numerically solve sustainable fashion supply chain game theory problems under oligopolistic competition and brand differentiation. There are two fashion firms, Firm 1 and Firm 2, each of which produces, stores, and distributes a single fashion product, which is differentiated by its brand. Each firm has, at its disposal, two manufacturing plants, two distribution centers, and serves a single demand market. The topology is illustrated in Fig. 6.2. The manufacturing plants M_1^1 and M_1^2 are located in the United States, whereas the manufacturing plants M_2^1 and M_2^2 are located offshore with lower operational costs. The demand market and the distribution centers are in the United States. The case study consists of three scenarios, each of which comprises a set of numerical examples.

We assume that each firm weights its total time with $\omega^1 = 1$, that is (cf. (6.7)), $\omega_1^1 = \omega_2^1 = 1$.

For the computation of solutions to the numerical examples, we implemented the Euler method, as discussed in Sect. 6.3, using MATLAB. The convergence tolerance was $\varepsilon = 10^{-6}$, and the sequence $\{a_\tau\} = .1(1, \frac{1}{2}, \frac{1}{2}, \ldots)$. We considered the algorithm to have converged (cf. (6.20)) when the absolute value of the difference between successive path flows differed by no more than the above ε. We initialized the algorithm by setting the demand of each fashion firm's product at 10 and equally distributed the demand among all the paths for each firm.

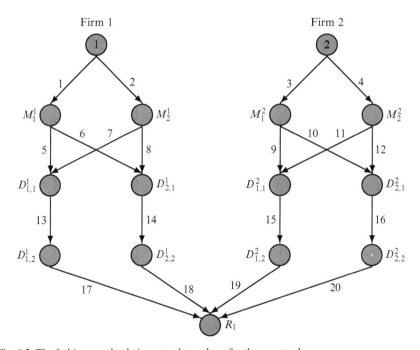

Fig. 6.2 The fashion supply chain network topology for the case study

Scenario 1. Fashion Firm 1 cares about the emissions that it generates much more than Firm 2 does, which is indicated, respectively, by $\omega_1^2 = 5$ and $\omega_2^2 = 1$. In addition, Firm 1 utilizes more advanced technologies in its supply chain activities in order to lower the emissions that it generates, but at relatively higher costs. The total generalized cost and the total emission functions are given in Table 6.1.

6.4 A Case Study

Table 6.1 Total generalized cost and total emission functions with link flow solution for Example 6.1

Link a	$\hat{g}_a(f)$	$\hat{e}_a(f_a)$	f_a^*	Link a	$\hat{g}_a(f)$	$\hat{e}_a(f_a)$	f_a^*
1	$10f_1^2+10f_1$	$.05f_1^2+.5f_1$	6.09	11	$1.5f_{11}^2+30f_{11}$	$.4f_{11}^2+1.5f_{11}$	9.28
2	$f_2^2+7f_2$	$.1f_2^2+.8f_2$	19.94	12	$1.5f_{12}^2+20f_{12}$	$.45f_{12}^2+f_{12}$	10.64
3	$10f_3^2+7f_3$	$.1f_3^2+f_3$	4.83	13	$f_{13}^2+3f_{13}$	$.01f_{13}^2+.1f_{13}$	11.48
4	$f_4^2+5f_4$	$.15f_4^2+1.2f_4$	19.93	14	$f_{14}^2+2f_{14}$	$.01f_{14}^2+.15f_{14}$	14.55
5	$f_5^2+4f_5$	$.08f_5^2+f_5$	2.85	15	$f_{15}^2+1.8f_{15}$	$.05f_{15}^2+.3f_{15}$	12.95
6	$f_6^2+6f_6$	$.1f_6^2+f_6$	3.24	16	$f_{16}^2+1.5f_{16}$	$.08f_{16}^2+.5f_{16}$	11.81
7	$2f_7^2+30f_7$	$.15f_7^2+1.2f_7$	8.63	17	$2f_{17}^2+f_{17}$	$.08f_{17}^2+f_{17}$	11.48
8	$2f_8^2+20f_8$	$.15f_8^2+f_8$	11.31	18	$f_{18}^2+4f_{18}$	$.1f_{18}^2+.8f_{18}$	14.55
9	$f_9^2+3f_9$	$.25f_9^2+f_9$	3.67	19	$f_{19}^2+5f_{19}$	$.3f_{19}^2+1.2f_{19}$	12.95
10	$f_{10}^2+4f_{10}$	$.25f_{10}^2+2f_{10}$	1.17	20	$1.5f_{20}^2+f_{20}$	$.35f_{20}^2+1.2f_{20}$	11.81

Example 6.1. At the demand market R_1, the consumers reveal their preferences for the product of Firm 1, through the demand functions, with the demand price functions for the two fashion products being given by

$$\rho_{11}(d) = -d_{11} - .2d_{21} + 300, \quad \rho_{21}(d) = -2d_{21} - .5d_{11} + 300.$$

The computed equilibrium link flows are reported in Table 6.1. For completeness, we also provide the computed equilibrium path flows in Table 6.2. Note that all the paths have positive flows and, hence, the *final* supply chain network topology in which links with positive equilibrium flow are depicted is as in Fig. 6.2.

Table 6.2 Computed equilibrium path flow pattern for Example 6.1

	Path definition	Path flow
O/D pair $w_1^1 = (1,R_1)$	$p_1 = (1,5,13,17)$	$x_{p_1}^* = 2.85$
	$p_2 = (1,6,14,18)$	$x_{p_2}^* = 3.24$
	$p_3 = (2,7,13,17)$	$x_{p_3}^* = 8.63$
	$p_4 = (2,8,14,18)$	$x_{p_4}^* = 11.31$
O/D pair $w_1^2 = (2,R_1)$	$p_1 = (3,9,15,19)$	$x_{p_1}^* = 3.67$
	$p_2 = (3,10,16,20)$	$x_{p_2}^* = 1.17$
	$p_3 = (4,11,15,19)$	$x_{p_3}^* = 9.28$
	$p_4 = (4,12,16,20)$	$x_{p_4}^* = 10.64$

The discrete-time trajectories of the path flows generated by the Euler method, as in (6.20), for Firm 1 are given in Fig. 6.3 and for Firm 2 in Fig. 6.4. The Euler method may be interpreted as a discrete-time tatonnement or adjustment process.

The computed equilibrium demands were

$$d_{11}^* = 26.03, \quad d_{21}^* = 24.76.$$

The incurred equilibrium prices at the demand markets were

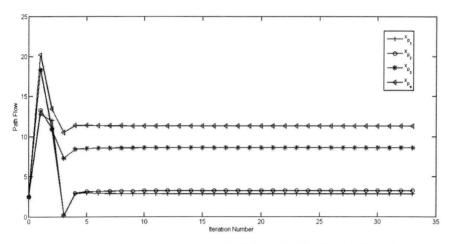

Fig. 6.3 Product path flow iterates generated by the Euler method for Firm 1 in Example 6.1

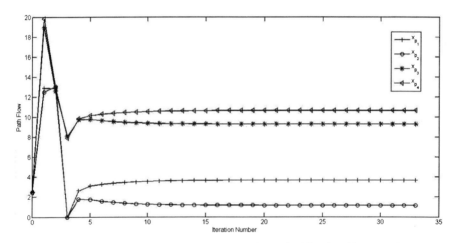

Fig. 6.4 Product path flow iterates generated by the Euler method for Firm 2 in Example 6.1

$$\rho_{11}(d^*) = 269.02, \quad \rho_{21}(d^*) = 237.47.$$

The total generalized cost for Firm 1 was 2,860.11; the total emissions that it generated 182.03; and its revenue was 7,002.35, yielding a profit of 4,142.25. The total cost for Firm 2 was 2,386.61; its total emissions 368.50; and its revenue 5,879.36, yielding a profit of 3,492.75. The utilities (cf. (6.11)) for Firm 1 and for Firm 2 were 3,232.11 and 3,124.25, respectively.

Example 6.2. Example 6.2 had the identical data to that of Example 6.1 except that the consumers were more price sensitive with respect to fashion Firm 2's product. The demand price function associated with Firm 2's product was now

6.4 A Case Study

$$\rho_{21}(d) = -3d_{21} - .5d_{11} + 300.$$

Example 6.3. Example 6.3 had the same data as Example 6.1, but now the consumers were even more price sensitive with respect to fashion Firm 2's product, with the demand price function for Firm 2's product now given by

$$\rho_{21}(d) = -4d_{21} - .5d_{11} + 300.$$

Example 6.4. Example 6.4 had the identical data as Example 6.1 except that the demand price function associated with fashion Firm 2's product was now

$$\rho_{21}(d) = -5d_{21} - .5d_{11} + 300.$$

The computed equilibrium demands, prices, profits, emissions, and utilities for Examples 6.1, 6.2, 6.3, and 6.4 are reported in Table 6.3.

We now provide some managerial insights from these examples. Note that the changes in the demand price function for fashion Firm 2's product (cf. Examples 6.1 through 6.4) indicate that the consumers are becoming more price sensitive with respect to fashion Firm 2's product. With the consumers' increasing environmental concerns, the demand for fashion Firm 2's product decreases significantly, since fashion Firm 2 does not have as good of a reputation in terms of environmental sustainability as Firm 1 does. In addition, the profit of fashion Firm 2 drops dramatically. The total emissions of Firm 1 increase slightly, whereas those of Firm 2 decrease substantially from Examples 6.1 through 6.4.

Table 6.3 Computed equilibrium demands, prices, profits, total emissions, and utilities for Examples 6.1, 6.2, 6.3, and 6.4

	Example 6.1	Example 6.2	Example 6.3	Example 6.4
Demand for Firm 1's product—d_{11}^*	26.03	26.12	26.18	26.22
Demand for Firm 2's product—d_{21}^*	24.76	20.68	17.75	15.55
Price of Firm 1's product—$\rho_{11}(d^*)$	269.02	269.75	270.27	270.67
Price of Firm 2's product—$\rho_{21}(d^*)$	237.47	224.91	215.91	209.14
Profit of Firm 1—π_1	4,142.25	4,168.39	4,187.19	4,201.36
Profit of Firm 2—π_2	3,492.75	2,881.09	2,451.92	2,134.32
Emissions of Firm 1—$\sum_{a \in L^1} \hat{e}_a(f_a^*)$	182.03	183.00	183.70	184.23
Emissions of Firm 2—$\sum_{a \in L^2} \hat{e}_a(f_a^*)$	368.50	269.75	208.22	167.04
Utility of Firm 1—$U_1(X^*)$	3,232.11	3,253.39	3,268.70	3,280.23
Utility of Firm 2—$U_2(X^*)$	3,124.25	2,611.34	2,243.70	1,967.27

Scenario 2. Scenario 2 consists of four examples. We assume that Firm 2 had become more environmentally conscious and raised ω_2^2 from 1 to 5. Hence, both firms have their ω^2 weights equal to 5. Examples 6.5 through 6.8 had their data identical to the data in Examples 6.1 through 6.4, respectively, except for the larger value of ω_2^2. The computed equilibrium demands, prices, profits, emissions, and utilities for Examples 6.5, 6.6, 6.7, and 6.8 are reported in Table 6.4.

Interestingly, the weights, the ω_i^2s, may also be interpreted as *taxes*. For example, a governmental authority may impose a tax associated with carbon emissions that each firm must pay. Hence, this set of examples in which the ω_i^2 terms are equal for both firms with a value of 5 reflects this policy option (see, e.g., Dhanda et al. 1999).

Table 6.4 Computed equilibrium demands, prices, profits, total emissions, and utilities for Examples 6.5, 6.6, 6.7, and 6.8

	Example 6.5	Example 6.6	Example 6.7	Example 6.8
Demand for Firm 1's product—d_{11}^*	26.18	26.23	26.27	26.30
Demand for Firm 2's product—d_{21}^*	17.35	15.13	13.41	12.04
Price of Firm 1's product—$p_{11}(d^*)$	270.34	270.74	271.05	271.30
Price of Firm 2's product—$p_{21}(d^*)$	252.20	241.50	233.23	226.65
Profit of Firm 1—π_1	4,189.75	4,204.08	4,215.16	4,224.00
Profit of Firm 2—π_2	3,019.58	2,563.82	2,225.83	1,965.67
Emissions of Firm 1—$\sum_{a \in L^1} \hat{e}_a(f_a^*)$	183.79	184.33	184.74	185.07
Emissions of Firm 2—$\sum_{a \in L^2} \hat{e}_a(f_a^*)$	191.27	152.71	125.73	106.03
Utility of Firm 1—$U_1(X^*)$	3,270.78	3,282.44	3,291.47	3,298.66
Utility of Firm 2—$U_2(X^*)$	2,063.20	1,800.29	1,597.17	1,435.52

Comparing the results of Examples 6.5, 6.6, 6.7, and 6.8 with those for Examples 6.1, 6.2, 6.3, and 6.4, respectively, we observe that, as expected, Firm 2 now emits a significantly lower amount, whereas Firm 1 now emits, in each example in this scenario, a slightly higher amount than it emitted in the corresponding example in Scenario 1. Nevertheless, an increase in environmental concerns is not sufficient for fashion Firm 2 to attract more demand and to increase its profit, since it has not modified its pollution-abatement technologies and, indeed, its total generalized cost functions and total emission functions remain as in Scenario 1. Such information is clearly useful to decision-makers and our theoretical and computational framework allows them to conduct sensitivity analysis to investigate the effects on profits and emissions of changes in the data.

Scenario 3. In our third and final scenario, we varied both the total generalized cost functions and the total emission functions of Firm 2 and also investigated the situation that the firms faced identical functions throughout the supply chain.

Example 6.9. Example 6.9 had the identical data to that in Example 6.5 except that Firm 2 now acquired more expensive advanced emission-reducing manufacturing technologies, resulting in new total generalized cost and emission functions associated with the manufacturing links, as below:

$$\hat{c}_3(f) = 10f_3^2 + 10f_3, \quad \hat{c}_4(f) = f_4^2 + 7f_4,$$
$$\hat{e}_3(f_3) = .05f_3^2 + .5f_3, \quad \hat{e}_4(f_4) = .1f_4^2 + .8f_4.$$

Example 6.10. Example 6.10 had the same data as Example 6.9, but now Firm 2 made even a greater effort to lower its emissions, not only focusing on its manufacturing processes but also on all other supply chain activities. The total generalized cost and the total emission functions are provided in Table 6.5.

6.4 A Case Study

Table 6.5 Total generalized cost and total emission functions for Example 6.10

Link a	$\hat{g}_a(f)$	$\hat{e}_a(f_a)$	Link a	$\hat{g}_a(f)$	$\hat{e}_a(f_a)$
1	$10f_1^2 + 10f_1$	$.05f_1^2 + .5f_1$	11	$2f_{11}^2 + 30f_{11}$	$.15f_{11}^2 + 1.2f_{11}$
2	$f_2^2 + 7f_2$	$.1f_2^2 + .8f_2$	12	$2f_{12}^2 + 20f_{12}$	$.15f_{12}^2 + f_{12}$
3	$10f_3^2 + 10f_3$	$.05f_3^2 + .5f_3$	13	$f_{13}^2 + 3f_{13}$	$.01f_{13}^2 + .1f_{13}$
4	$f_4^2 + 7f_4$	$.1f_4^2 + .8f_4$	14	$f_{14}^2 + 2f_{14}$	$.01f_{14}^2 + .15f_{14}$
5	$f_5^2 + 4f_5$	$.08f_5^2 + f_5$	15	$f_{15}^2 + 3f_{15}$	$.01f_{15}^2 + .1f_{15}$
6	$f_6^2 + 6f_6$	$.1f_6^2 + f_6$	16	$f_{16}^2 + 2f_{16}$	$.01f_{16}^2 + .15f_{16}$
7	$2f_7^2 + 30f_7$	$.15f_7^2 + 1.2f_7$	17	$2f_{17}^2 + f_{17}$	$.08f_{17}^2 + f_{17}$
8	$2f_8^2 + 20f_8$	$.15f_8^2 + f_8$	18	$f_{18}^2 + 4f_{18}$	$.1f_{18}^2 + .8f_{18}$
9	$f_9^2 + 4f_9$	$.08f_9^2 + f_9$	19	$2f_{19}^2 + f_{19}$	$.08f_{19}^2 + f_{19}$
10	$f_{10}^2 + 6f_{10}$	$.1f_{10}^2 + f_{10}$	20	$f_{20}^2 + 4f_{20}$	$.1f_{20}^2 + .8f_{20}$

Example 6.11. Example 6.11 had the identical data as in Example 6.10 except that the effort made by Firm 2 to protect the environment was now also disseminated to the consumers, leading to the change in the demand for Firm 2's product, with the new demand price function faced by Firm 2 given by

$$\rho_{21}(d) = -d_{21} - .2d_{11} + 300.$$

The computed equilibrium demands, prices, profits, emissions, and utilities for Examples 6.1, 6.5, 6.9, 6.10, and 6.11 are given in Table 6.6.

Table 6.6 Computed equilibrium demands, prices, profits, total emissions, and utilities for Examples 6.1, 6.5, 6.9, 6.10, and 6.11

	Example 6.1	Example 6.5	Example 6.9	Example 6.10	Example 6.11
Demand for Firm 1's product—d_{11}^*	26.03	26.18	26.18	26.11	26.00
Demand for Firm 2's product—d_{21}^*	24.76	17.35	17.75	20.80	26.00
Price of Firm 1's product—$\rho_{11}(d^*)$	269.02	270.34	270.27	269.73	268.80
Price of Firm 2's product—$\rho_{21}(d^*)$	237.47	252.20	251.40	245.34	268.80
Profit of Firm 1—π_1	4,142.25	4,189.75	4,187.18	4,167.60	4,134.29
Profit of Firm 2—π_2	3,492.75	3,019.58	3,020.61	3,137.78	4,134.29
Emissions of Firm 1 —$\sum_{a \in L^1} \hat{e}_a(f_a^*)$	182.03	183.79	183.70	182.97	181.73
Emissions of Firm 2 —$\sum_{a \in L^2} \hat{e}_a(f_a^*)$	368.50	191.27	182.55	127.73	181.73
Utility of Firm 1—$U_1(X^*)$	3,232.11	3,270.78	3,268.68	3,252.75	3,225.64
Utility of Firm 2—$U_2(X^*)$	3,124.25	2,063.20	2,107.85	2,500.92	3,225.64

Based on the results for Examples 6.5, 6.9, and 6.10, the advanced manufacturing technologies utilized by fashion Firm 2 did improve its performance, but not significantly, while Firm 2's environmental efforts throughout its supply chain notably enhanced its profit and utility. Furthermore, and, this is relevant also to managers, the change in consumers' attitudes towards Firm 2 can assist Firm 2 in obtaining as much profit as that of Firm 1. Observe that the profit of Firm 1 in Example 6.11, however, was not as high as what it achieved in Example 6.1, which means that if Firm 1 wishes to maintain its competitive advantage, it must pay continuing attention to its emissions. A comparison of the results in Example 6.10 and Example 6.11, in turn, suggests that the development of a positive image for a firm in terms of its environmental consciousness and concern may also be an effective marketing strategy for fashion firms.

The above case study demonstrates that consumers' environmental consciousness can be a valuable incentive to spur fashion companies to reexamine their supply chains so as to reduce their environmental pollution, which can, in turn, help such companies to obtain competitive advantages and increased profits.

6.5 Summary and Conclusions

In this chapter, we focused on the fashion and apparel industry, which presents unique challenges and opportunities in terms of environmental sustainability. We developed a competitive supply chain network model, using variational inequality theory, that captures oligopolistic competition with fashion product brand differentiation. The variational inequality model assumes that each firm seeks to maximize its profit, which includes generalized costs that capture supply chain activity time consumption, and to minimize the emissions that it generates throughout its supply chain as it engages in its activities of manufacturing, storage, and distribution, with a weight associated with the latter criterion. The model allows for alternative modes of transportation from manufacturing sites to distribution centers and from distribution centers to the demand markets, since different modes of transportation are known to emit different amounts of emissions.

The variational inequality-based competitive supply chain network model advances the state-of-the-art of supply chain modeling in several ways:

- It captures competition through brand differentiation, which is an important feature of the fashion industry.
- It allows for each firm to individually weight its concern for the environment in its decision-making.
- It utilizes time consumption functions associated with the supply chain network activities to allow for operational cost versus time trade-offs.
- It includes alternatives such as multiple modes of transportation.

In order to demonstrate the generality of the model and the proposed computational scheme, we presented a case study, in which, through a series of numerical examples, we demonstrated the effects of changes on the demand price functions, the total generalized cost and total emission functions, as well as the weights associated with the environmental criterion on the equilibrium product demands, the product prices, profits, and utilities.

The case study also demonstrated that consumers can have a major impact, through their environmental consciousness, on the level of profits of firms in their favoring of firms that adopt environmental pollution-abatement technologies for their supply chain activities. The numerical examples in the case studies were selected for their transparency and for reproducibility purposes.

6.6 Sources and Notes

This chapter is based on the paper by Nagurney and Yu (2012). However, here, we assume that the link costs are *generalized* link costs that include not only the operational costs but also the monetarized value of time, with explicit link time consumption functions associated with the supply chain network activities. Hence, the scope of the model is broader than that in Nagurney and Yu (2012). In addition, we report here expanded case study output results. A related paper is that of Nagurney and Yu (2011), which focused on multicriteria decision-making for fashion supply chain management with the minimization of cost (see also Nagurney 2010) and the minimization of time as relevant criteria and assumed that the demand for the fashion product was known at each demand market. The model in this chapter, in contrast, assumes that the demand for the particular brand is elastic and not fixed and considers multiple, competing firms rather than a single firm. For an edited volume on fashion supply chain management, a relatively new area of application of rigorous tools, see Choi (2011), the focused journal special issue edited by Choi and Chen (2008), and the papers by Sen (2008) and Brun et al. (2008).

This chapter builds on the existing literature in sustainable supply chain management, with a focus on system-wide issues and in an industry in which competition and brands are the reality. Indeed, as early as Beamon (1999), Sarkis (2003), Corbett and Kleindorfer (2003), Nagurney and Toyasaki (2003), Sheu et al. (2005), Kleindorfer et al. (2005), and Linton et al. (2007), it has been argued that sustainable supply chains are critical for the examination of operations and the environment, with sustainable fashion being a more recent topic in both research and practice (see, e.g., de Brito et al. 2008 and the references therein). Sustainable supply chains have arisen as a focus for special issues (see Piplani et al. 2008) and have advanced to a degree that even policies to reduce emissions have been explored in rigorous frameworks (see Nagurney et al. 2006 and Chaabane et al. 2012). For a thorough survey of sustainable supply chain management until 2008, see Seuring and Muller (2008). Nevertheless, a general, rigorous modeling and computational framework that captures oligopolistic competition, brand differentiation, and environmental concerns, in a supply chain network setting has not, heretofore, been constructed.

We now, for completeness, discuss another related issue of global supply chains, of relevance to various industrial sectors, from fast fashion to high tech—that of global outsourcing and quick-response production. Specifically, we note that global outsourcing and quick-response production decisions of supply chain firms can be captured in a network perspective (cf. Fig. 6.5). As noted in Liu and Nagurney (2012), one can allow multiple suppliers, multiple manufacturers, and multiple demand markets to interact under both demand and cost uncertainty. In particular, each manufacturer maximizes its own expected profit through a two-stage stochastic programming problem, while competing with the other manufacturers, but cooperates with the offshore suppliers in the first stage. The governing equilibrium conditions of the entire supply chain network are formulated as a variational inequality formulation.

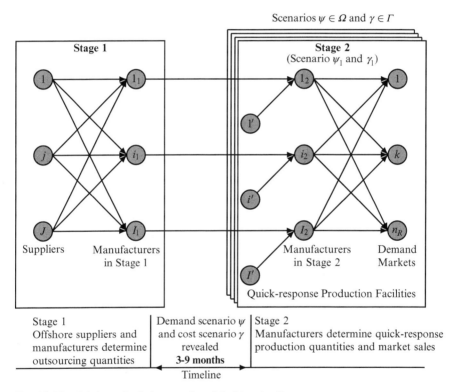

Fig. 6.5 The global supply chain network and decision timeline

We emphasize that the model in this chapter is not limited to fast fashion apparel supply chains. Since it is a competitive oligopoly supply chain network model, it is also relevant to numerous other oligopolies that underpin our economy—from computer and high tech manufacturing companies to many utility and energy companies to wireless service providers to airlines and airplane manufacturers, as well as automobile companies, to even certain banks. Of course, the criterion of emission minimization may be replaced with another criterion, if appropriate, such as, risk minimization, in the case of banks, or the associated weight set to zero. The notable feature of the model in this chapter, as of the models in Part III of this book, is the explicit incorporation of product/brand differentiation, which enables consumers to differentiate the products, based on the manufacturers/producers, who, in our analytics framework, are represented by their supply chain networks. Hence, quality is implicit in our models and time a fundamental characteristic and feature, whether captured through arc multipliers and links, as in the generalized network models in Chaps. 4 and 5, or through the objective function, as in this chapter (and on a different time scale, through the algorithm/iterative procedures that track the dynamic adjustment processes of the firms until an equilibrium is achieved). Whether in food or pharmaceuticals and healthcare, in fast fashion or high tech, and even the Internet, those firms that optimize their supply chain networks to compete with and against time will become the winning time-sensitive and perishable product and service brands.

References

Beamon B (1999) Designing the green supply chain. Logist Inform Manag 12: 332–342

Brun A, Caniato F, Caridi M, Castelli C, Miragliotta G, Romchi S, Sianesi A, Spina G (2008) Logistics and supply chain management in luxury fashion retail: empirical investigation of Italian firms. Int J Prod Econ 114(2):554–570

Carlton D, Perloff J (2004) Modern industrial organization, 4th edn. Pearson-Addison Wesley, Reading, MA

Chaabane A, Ramudhin A, Paquet M (2012) Design of sustainable supply chains under the emission trading scheme. Int J Prod Econ 135(1):37–49

Choi T-M (ed) (2011) Fashion supply chain management: industry and business analysis. IGI Global, Hershey, Pennsylvania

Choi T-M, Chen Y (2008). Editorial. Int J Prod Econ 114(2):415

Claudio L (2007) Waste couture: Environmental impact of the clothing industry. Environ Health Perspect 115:449–454

Corbett CJ, Kleindorfer PR (2003) Environmental management and operations management: Introduction to the third special issue. Prod Oper Manag 12(3):287–289

de Brito MP, Carbone V, Meunier Blanquart C (2008) Towards a sustainable fashion retail supply chain in Europe: organisation and performance. Int J Prod Econ 114(2):534–553

Dhanda KK, Nagurney A, Ramanujam P (1999) Environmental networks: a framework for economic decision-making and policy analysis. Edward Elgar Publishing, Cheltenham

Doeringer P, Crean S (2006) Can fast fashion save the U.S. apparel industry? Soc Econ Rev 4(3):353–377

Gereffi G, Memedovic O (2003) The global apparel value chain: What prospects for upgrading by developing countries. Sectoral Studies Series, United Nations Industrial Development Organization, Vienna, Austria

Glausiusz J (2008) Eco-chic to the rescue! Discover Magazine, September

Guardian (2010) Company profile for H&M. Available online at: http://wwwguardian.co.uk/sustainable-business/companies-list-h-i1

Kleindorfer PR, Singhal K, Van Wassenhove LN (2005) Sustainable operations management. Prod Oper Manag 14(4):482–492

Linton JD, Klassen R, Jayaraman V (2007) Sustainable supply chains: an introduction. J Oper Manag 25(6):1075–1082

Liu Z, Nagurney A (2012) Supply chain networks with global outsourcing and quick-response production under demand and cost uncertainty. Ann Oper Res, in press. DOI: 10.1007/s10479-011-1006-0

Nagurney A (2010) Optimal supply chain network design and redesign at minimal total cost and with demand satisfaction. Int J Prod Econ 128(1):200–208

Nagurney A, Liu Z, Woolley T (2006) Optimal endogenous carbon taxes for electric power supply chains with power plants. Mathe Comput Model 44(9–10):899–916

Nagurney A, Toyasaki F (2003) Supply chain supernetworks and environmental criteria. Transport Res D 8(3):185–213

Nagurney A, Woolley T (2010) Environmental and cost synergy in supply chain network integration in mergers and acquisitions. In: Ehrgott M, Naujoks B, Stewart T, Wallenius J (eds) Sustainable energy and transportation systems, Proceedings of the 19th international conference on multiple criteria decision making, Lecture Notes in Economics and Mathematical Systems. Springer, Berlin, pp 51–78

Nagurney A, Yu M (2011) Fashion supply chain management through cost and time minimization from a network perspective. In: Choi T-M (ed) Fashion supply chain management: industry and business analysis. IGI Global, Hershey, Pennsylvania, pp 1–20

Nagurney A, Yu M (2012) Sustainable fashion supply chain management under oligopolistic competition and brand differentiation. Int J Prod Econ 135(2): 532–540

Nordas HK, Pinali E, Geloso Grosso M (2006) Logistics and time as a trade barrier. OECD Trade Policy Working Papers, No. 35, OECD Publishing. doi:10.1787/664220308873

Piplani R, Pujawan N, Ray S (2008) Foreword: sustainable supply chain management. Int J Prod Econ 111(2):193–194

Rabon L (2001) Technology outlook 2001: U.S. suppliers weather winds of change. Bobbin 42:54–60

Rosenbloom S (2010) Fashion tries on zero waste design. The New York Times, August 13

Sarkis J (2003) A strategic decision framework for green supply chain management. J Cleaner Prod 11(4):397–409

Sen A (2008) The US fashion industry: A supply chain review. Int J Prod Econ 114(2):571–593

Seuring S, Muller M (2008) From a literature review to a conceptual framework for sustainable supply chain management. J Cleaner Prod 16(15):1699–1710

Sheu J-B, Chou Y-H, Hu C-C (2005) An integrated logistics operational model for green-supply chain management. Transport Res E 41(4):287–313

Shy O (1996) Industrial organization: theory and applications. MIT Press, Cambridge, MA

So KC (2000) Price and time competition for service delivery. Manuf Serv Oper Manag 2(4):392–409

Sull D, Turconi S (2008) Fast fashion lessons. Bus Strat Rev Summer 5–11

Tucker R (2010) NRDC tackles China's textile pollution. Women's Wear Daily, June 29

UPI.com (2010) Jeans add to Pearl River's pollution woes. Available online at: http://www.upi.com/Science_News/Resource_Wars/2010/04/29/Jeans-add-to-Pearl-Rivers-pollution-woes/UPI-37351272566965/

World Resources Institute (1998) World Resources 1998–99: environmental change and human health. Oxford University Press, Oxford

Glossary of Notation

This is a glossary of symbols and basic definitions useful in understanding this book. Other symbols and terms are defined in the book as needed. A vector is assumed to be a column vector, unless noted otherwise.

$\in$	An element of
$\subset$	Subset of
$\subseteq$	Subset of or equal to
$\forall$	For all
$\exists$	There exists
R	The real line
R^n	Euclidean n-dimensional space
R^n_+	Euclidean n-dimensional space on the nonnegative orthant
:	Such that; also $\vert$
$\equiv$	Is equivalent to
$\mapsto$	Maps to
$\rightarrow$	Tends to
$\|x\| = (\sum_{i=1}^n x_i^2)^{\frac{1}{2}}$	Length of $x \in R^n$ with components $(x_1, x_2, \ldots, x_n)$
x^T	Transpose of a vector x
$\langle x, x \rangle$	Inner product of vector x in R^n where $\langle x, x \rangle = x_1^2 + \ldots + x_n^2$
$x^T \cdot x$	Also denotes the inner product of x
$\vert y \vert$	Absolute value of y
$[a,b]; (a,b)$	A closed interval; an open interval in R
∇f	Gradient of $f : R^n \mapsto R$
∇F	The $n \times n$ Jacobian of a mapping $F : R^n \mapsto R^n$
$\frac{\partial f}{\partial x}$	Partial derivative of f with respect to x
A^T	Transpose of the matrix A
$\prod_{i=1}^n X_i$	Cartesian product of sets $X_1 \times X_2 \times \ldots \times X_n = \{(x_1, \ldots, x_n) : x_i \in X_i\}$
$\operatorname{argmin}_{x \in K} f(x)$	The set of $x \in K$ attaining the minimum of $f(x)$

Convex set	A set $S \subseteq R^n$ is convex if and only if $\forall x, y \in S$ and $\lambda \in [0, 1]$, we have that $\lambda x + (1 - \lambda) y \in S$.
Convex function	Let $f : S \subseteq R^n \mapsto R$, and S be a convex set. The function f is convex if and only if $\forall x, y \in S$ and $\forall \lambda \in (0, 1)$: $f(\lambda x + (1 - \lambda) y) \leq \lambda f(x) + (1 - \lambda) f(y)$.